SELF-EVIDENT ASTROLOGY™

Jeffrey Sayer Close

ISBN-10: 0-86690-592-8
ISBN-13: 978-0-86690-592-3

Cover Design: Jack Cipolla
Editing: Stephanie Jean Clement Ph.D.

Published by:
American Federation of Astrologers, Inc.
6535 S. Rural Road
Tempe, AZ 85283

www.astrologers.com

Printed in the United States of America

Contents

Introduction   v

The Sun, Moon and Eight Planets   1

Symbols of the Planets Decoded   25

The Eight Planetary Companions   35

Further Evidence and Comments on Use in Interpretation   53

Mathematics, the Solar System and Rulerships   63

Appendix I, Planet Data   81

Appendix II, Planetary Moon Data   87

Illustrations   93

Introduction

SELF-EVIDENT ASTROLOGY™ is a new approach to astrology. Unlike the many texts that further refine the empirical traditions of Western and Eastern astrology, this book takes a different tack. SELF-EVIDENT ASTROLOGY™ is more basic and fundamental than any previous school of thought in astrology.

SELF-EVIDENT ASTROLOGY™ is a new view of all facets of astrology. It is the only approach that unifies all astrology inside a single common framework, and could be called the first true theory of astrology. Lastly, SELF- EVIDENT ASTROLOGY™ reintegrates astrology and astronomy.

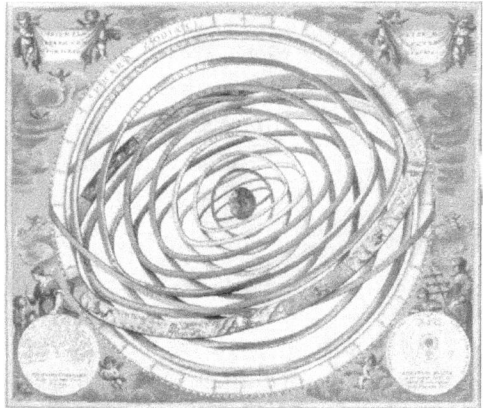

In most fields of study, one makes observations. After enough observations one begins to see patterns. These patterns suggest a logic behind the patterns that forms a theory. To find out if the theory is true, one tests the theory and makes more observations.

This balance of theory and observations is common to most every field of study, but is not clearly evident in astrology. Until now, astrology has been almost entirely an empirical field. Observations have been made correlating the positions of the planets and human behavior. While some generalities have been made from these observations, there has never been a comprehensive unified theory of astrology.

SELF-EVIDENT ASTROLOGY™ now presents such a theory. Like any good theory, it takes a complicated subject and shows that it has simple fundamentals that anyone can understand. It should be a simpler and easier approach to astrology. Most traditional interpretations of astrology are not contradicted by SELF-EVIDENT ASTROLOGY™. In fact, most are very much the same. There are places where SELF-EVIDENT ASTROLOGY™ further refines or adjusts traditional interpretations. SELF-EVIDENT ASTROLOGY™ also extends our understanding of many facets of astrology:

- Meanings of nineteen planetary moons.
- Five new progressions.
- The Eight Planetary Companions.
- A revision of the rulerships (association of planets and zodiac signs).

SELF-EVIDENT ASTROLOGY™ is self-evident. What could be more natural than to suggest that the planets mean what they are and that they are what they mean? More specifically, SELF-EVIDENT ASTROLOGY™ will show that the meanings of the eight planets in our sky form this simple matrix:

	Integration	Separation
Individual	Venus	Mars
Family	Saturn	Jupiter
Community	Uranus	Neptune
Humanity	Mercury	Pluto

Venus, for example, means the integration (coming together) of two individuals. Pluto means separated from all humanity. Saturn is concerned with family integration. These statements may seems too simple; however, for a theory to be truly fundamental, it should be simple.

SELF-EVIDENT ASTROLOGY™ assumes that any fundamental theory should break a subject down into its simple component pieces, often groupings of two, three or four as shown here.

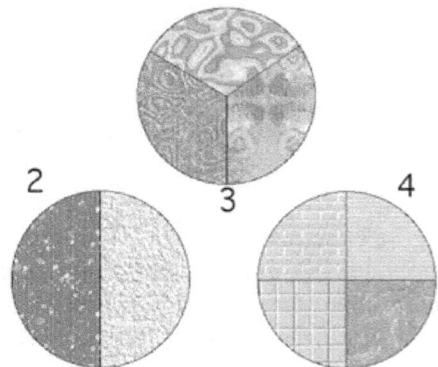

Throughout our daily lives we see life broken down into small pieces. We have two genders: male and female. We have three stages of life: child, parent and grandparent. The computer monitors and TV screens we have been watching for years have their basis in a three-color system (Red Green Blue, commonly known as "RGB"). Publishers and printing companies often use a four-color system. This is necessary as the printer is working from a white background, whereas the TV or computer monitor is working from a black background. (Cyan Magenta Yellow Black, aka CMYK.)

From ancient times we have the dual concepts of yin and yang. Every year we have four seasons. We have two solstices and two equinoxes. There are four directions on the face of a compass. Our location on Earth is expressed as a combination of latitude and longitude. In the USA and many other countries there are three branches of government.

The point is simple: our lives seem to break down into groups of two, three and four, etc. Astrology has traditionally been broken into small parts—often two, three or four of something. While we have twelve signs of the zodiac, we have divisions of three and four. The four types of signs are earth, water, air and fire. The signs also break into three groups, cardinal, fixed and mutable.

Aries is a cardinal fire sign, but just what does that mean? Isn't there a better way to express these groupings? The problem with traditional astrology is that there is no objective rationale behind why the groupings exist. In SELF-EVIDENT ASTROLOGY™ we look for the basis of the groupings. In order to find this basis, the first step is to take a look at some of the obvious groupings that occur in groups, such as two, three or four.

Many have looked into astrology but not pursued its study because it is was so mysterious and complicated. SELF-EVIDENT ASTROLOGY™ ends the confusion by reuniting astronomy and astrology.

In SELF-EVIDENT ASTROLOGY™ the planets mean what they are and are what they mean. It is this simple. All we need in order to understand astrology is to look into the nature of the solar system and sky around us. This book will show that the physical arrangement of the solar system, the orbital positions, the inclinations of orbital planes and things such as the mere size of a planet lead to a natural meaning for each planet.

If, as SELF-EVIDENT ASTROLOGY™ suggests, the meaning of the planets breaks down into the following simple pattern:

	Integration	Separation
Individual	Venus	Mars
Family	Saturn	Jupiter
Community	Uranus	Neptune
Humanity	Mercury	Pluto

then the implications are far-reaching.

Jeffrey Sayer Close

Chapter One

The Sun, Moon and Eight Planets

Our solar system has the Sun, our Moon and eight[1] planets aside from Earth. There are also a great number of asteroids and other moons. To get started in the solar system we will limit discussion to the Sun and Moon. Let's take a look at the physical characteristics of these two bodies and see what astrological meanings they suggest.

The kingpin of the solar system is the Sun. The Sun contains 99.85% of the entire mass of our solar system and Jupiter has most of the rest. We all know that life would not exist without the Sun. The Sun constantly gives off energy equal to 383 billion trillion kilowatts (383,000,000,000,000,000,000,000,000 watts). (See Figure 1.)

Not only is the Sun the source of the daily sunlight needed for virtually all life to grow, the energy of the Sun has been stored in the form of fossil fuels over millions of years, so the Sun is the source of fossil fuel energy as well.

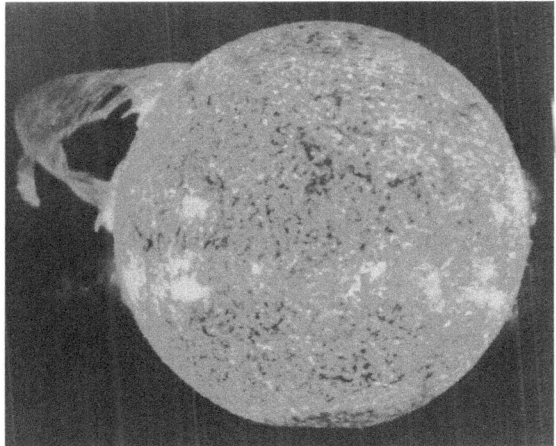

Figure 1. The Sun.

---

[1]The International Astronomical Union has demoted Pluto from the status of planet. Neither astrologers in general nor this author agree with this designation.

The Sun is the source of virtually all energy in the solar system[2] Hence, it would seem to represent the source of all things. The Sun's symbol (see Figure 2), the circle, includes a point at the center that radiates out in all directions. This symbol exists in the art of cultures all over the planet. The circle with a center is the most basic mandala and symbol for wholeness. From the human perspective, then, physical qualities of the Sun suggest the Sun in the astrological chart also represents the source of existence for all persons.

The Sun dwarfs all of our planets put together. Hence it seems be safe to assume that the meaning of the Sun is something greater than the any one planet. Perhaps, in some ways, is the Sun more important than the planets?

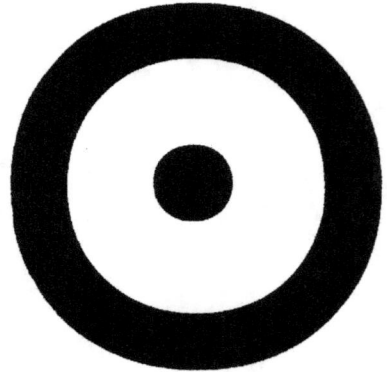

Figure 2. Symbol for the Sun.

Did you ever stop to think how it is that the Moon is just the right size to just barely cover the Sun during a total solar eclipse? Oddly enough, the Moon is 400 times smaller than the Sun, but also just happens to be 400 times nearer to the Earth. In fact, many millions of years ago the Moon appeared larger than the Sun because it was actually closer to the Earth than it is today[3]. The fact that the Sun and Moon appear equal in the sky is reflected in astrology by the fact that the Sun and Moon are the prime movers in any chart.

For us on Earth, the Moon appears to be the equal of the Sun. The Sun rules the day, but the Moon rules the night.

The Moon's symbol is composed of two semi-circles and looks like a satellite receiving dish—a parabolic surface that focuses incoming energy as a single destination. The Moon thus represents where all things end, where they come to completion.

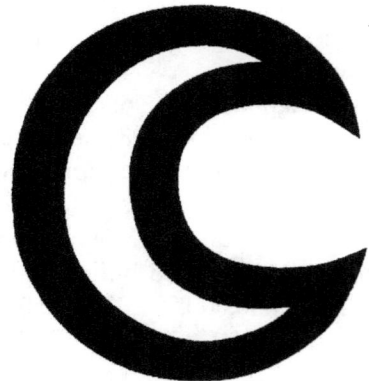

Figure 3. Symbol for the Moon.

The Moon reflects the Sun's light and the Moon is most visible during the time of day (nighttime) when we reflect upon the day's events. Whether we are out sitting on a porch in the late evening or taking our night's sleep, this is a time when we reflect on what has happened during the day.

---

[2]Jupiter and some other outer planets actually radiate more energy than they absorb, but the amounts are extremely small compared to the Sun's emissions.

[3]Over time the Moon's orbit is moving further and further from Earth at the current average of 3.8 cm per year. The Moon is slightly egg-shaped and the small end of the Moon is always facing the Earth. This is why the rotation of the Moon is synchronized to its orbit and the same side of the Moon always faces the Earth.

Figure 4. The Moon.

At the Full Moon, the Moon provides us with enough reflected energy to actually see shadows. We have shadows during the day, but we don't think much about them. At night, shadows are sometimes all we can see. Shadows, a reflection, are a characteristic of how we view the Moon (see Figure 4).

Even water, another item related to reflection, interacts with the Moon. The interaction with the Earth's tides is the primary reason for the exceptionally complicated orbit of the Moon. And these tides help give life to the sea. Many creatures have biological events timed to the phases of the Moon.

Looking to the distant past, a few billion years ago, the Moon was closer to Earth and the tides more pronounced. Some astronomers theorize that the interaction of the Moon and the large tides helped to "mix the primordial soup" that became life on Earth.

The Sun and Moon in Astrological Charts

Let's look at how the Sun and Moon are placed in the birth charts of George W. Bush and John Lennon in order to illustrate the Sun and Moon positions and look at the interpretation under SELF-EVIDENT ASTROLOGY™. The meanings of zodiac signs and house placements are from portions of SELF-EVIDENT ASTROLOGY™ not covered in this book.

The Sun is at 13° Cancer 47′ in the twelfth house in George W. Bush's chart (see Figure 5).

Sun	Source, point of radiation, outflow
Cancer	Starting family
Twelfth House	Self-separated from community—Endings

Put altogether, these considerations suggest that George W. Bush instigates actions (Sun) that, taken together (Cancer), form a new family of endings (twelfth house).

The Moon in Bush's chart is at 16° Libra 59′ in the third house (see Figure 5).

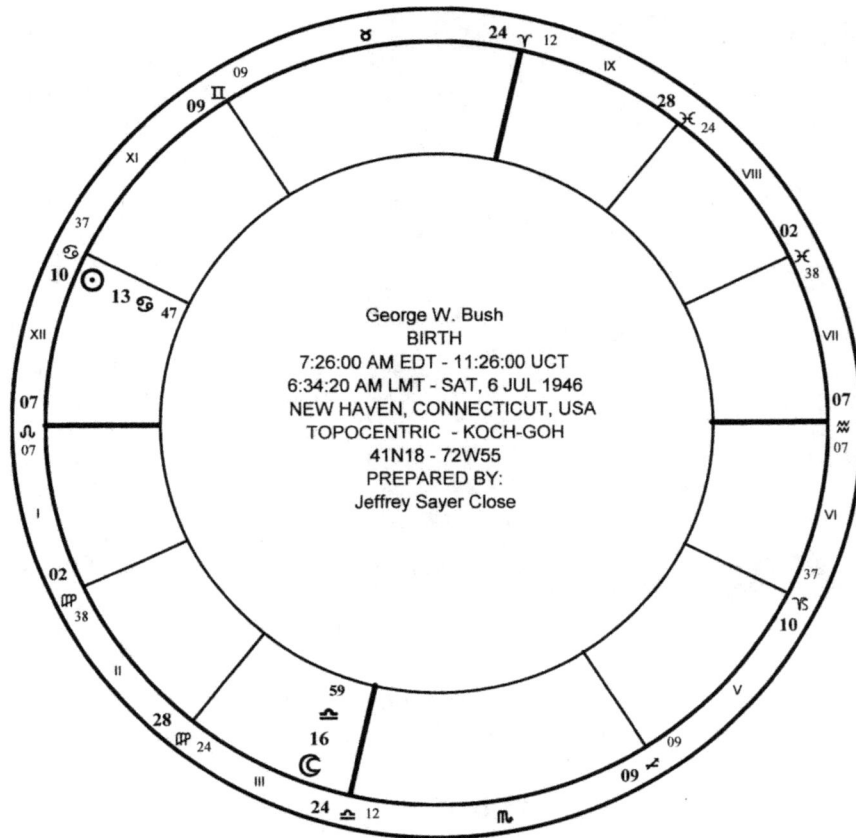

Figure 5. Georege W. Bush, Sun and Moon.

Moon	Closure, point of reception, inflow
Libra	Starting of others, of humanity
Third House	Self-integrated into a family—learning

Put together, these meanings suggest that George W. Bush pulls in (Moon) learning (information) (third house) from all humanity (Libra).

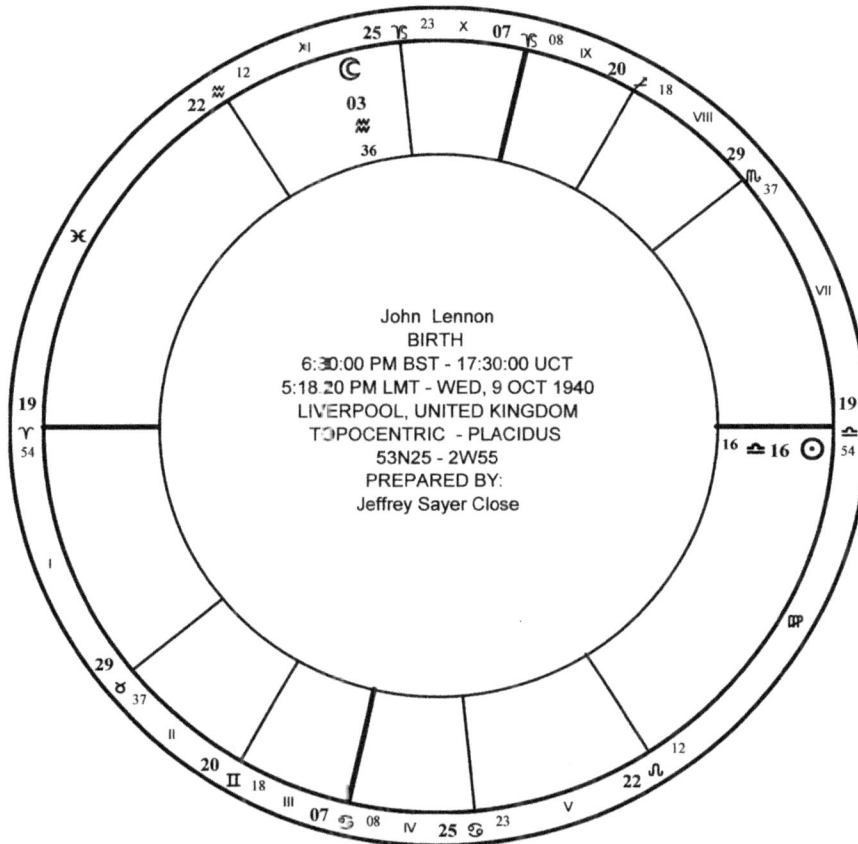

Figure 6. John Lennon, Sun and Moon.

The Sun in John Lennon's chart is at 16° Libra 16′ in the sixth house (see Figure 6).

Sun	Source, point of radiation, outflow
Libra	Starting of others, of humanity
Sixth House	Others separated from family—Health and perfection

John Lennon worked throughout his life to inspire all of humanity to perfection.

The Moon in Lennon's chart is at 3° Aquarius 36′ in the eleventh house (see Figure 6).

Moon	Closure, point of reception, inflow
Aquarius	Continuation of others, of humanity
Eleventh House	Self-integrated into community—Career and goals

Lennon brought a community of humanitarians into his life.

How the Planets Fit into the Astrological Picture

If you look at the Sun and Moon, you have the Sun starting all things, and then all things ending with the Moon. The eight planets modify all things as shown below:

Start	Change/Continue	End/Close
	Mercury	
	Venus	
	Mars	
Sun	Jupiter	Moon
	Saturn	
	Uranus	
	Neptune	
	Pluto	

Note that the Sun and Moon are in essentially circular orbits that remain constant. The eight planets move at different speeds and even appear to move backwards (retrograde) in the sky at times. Hence the eight planets have orbits that appear to change, but the Sun and Moon do not.

If you look at the solar system from above (see Figure 7), all of the planets appear to move forward in more or less regular orbits, and none ever appear to go retrograde. Retrogradation is a phenomenon that only has meaning in the solar system when viewed from one of the planets. The Sun and Moon don't appear to change their orbits, but all eight of the other planets appear to reverse direction periodically. Hence we have one of our simple three part fundamentals: Sun is "start," the eight planets are "change" and the Moon is "end."

Moving beyond just the Sun and the Moon, the question is what type of change do each of the eight planets represent?

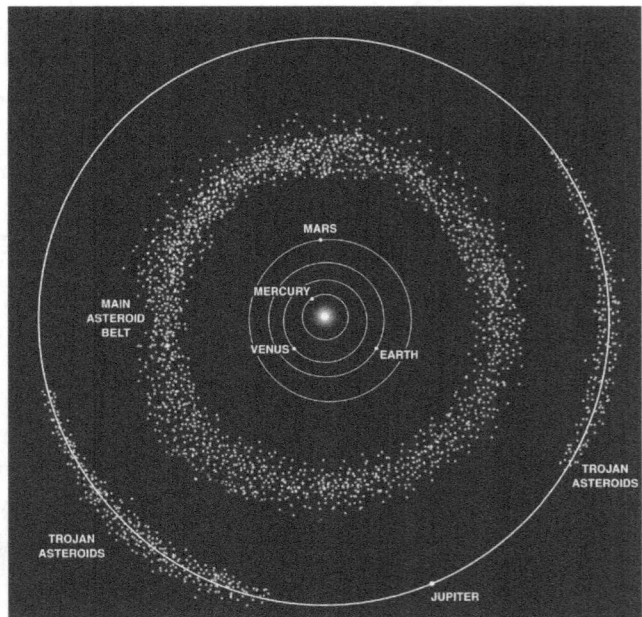

Figure 7. Solar System.

Let's take a basic approach and examine the eight planets as a group to see what the most obvious (self-evident) features of the solar system are.

In Figure 7, the most obvious feature is that there is an asteroid belt dividing the small planets from the larger ones. This Mars-Jupiter asteroid belt forms a division between the small inner planets of Mercury, Venus, Earth and Mars and the large outer planets Jupiter, Saturn, Uranus, Neptune and Pluto. (Pluto is actually smaller than any of the inner planets.) Figure 7 also shows the planetary orbits from Mercury out to Jupiter. Beyond Jupiter are the orbits of Saturn, Uranus, Neptune and Pluto. The point of the diagram is to show the location of the Mars-Jupiter asteroid belt (ring of dots).

One way to look at the difference between the inner planets and most of the outer planets is that the four inner planets (Mercury, Venus, Earth and Mars) have similar diameters. Figure 8 shows the relative diameters of the planets to help illustrate planetary pairings by planetary radius.

Ignoring Pluto for the moment, the four outer planets (Jupiter, Saturn, Uranus and Neptune) also have large diameters. The difference in the average size of the outer planets relative to the inner planets is roughly an order of magnitude or ten times greater. For detailed information on the physical characteristics of the planets see Appendix I.

When you view an outer planet with its large and numerous moons what does it remind you of? Most people would answer that it looks like a parent with a group of children—or, for short, a family.

In the table on page 8, "inner" and "outer" refer to the planetary position with respect to the Mars-Jupiter Asteroid Belt. The words "single" and "family" refer to the fact that the "single" planets are small and have very few or no moons. The "family" planets are large planets with numerous moons,

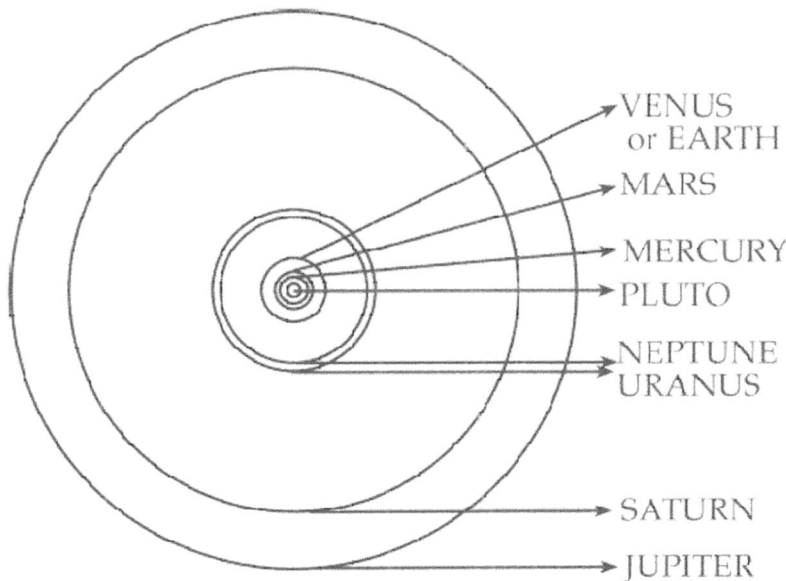

Figure 8. Relative Diameters of the Planets.

VENUS or EARTH
MARS
MERCURY
PLUTO
NEPTUNE
URANUS
SATURN
JUPITER

some the size of some of the "inner" planets, hence suggesting a "family" or group. Let's look at a chart[4]:

Planet	Moons	Rings	Type
Mercury	-	0	Inner, Single
Venus	-	0	Inner, Single
Earth	1	0	Inner, Single
Mars	2	0	Inner, Single
(Asteroid belt)			
Jupiter	49	4	Family/Outer
Saturn	52	7	Family/Outer
Uranus	27	11	Family/Outer
Neptune	13	4	Family/Outer
Pluto	3	0	Single/Outer

Figure 9 is a comparison of most of the largest of the planetary moons. With Earth as a size reference, you can see that the larger moons are comparable in size to Mars, Mercury and Pluto.

The next most obvious characteristics of the planets are their size and position of their orbits as you move out from the Sun. Both of these characteristics are shown in Figure 10. (For more details on orbits and other characteristics see Appendix I.) From left to right you have the Sun, Mercury, Venus, Mars and then the much larger Jupiter, Saturn, Uranus, Neptune and lastly the tiny Pluto.

In the introduction, I suggested that breaking down complex systems can sometimes be achieved by looking at groupings of two, three, or four of something where there is an element of commonality. In SELF-EVIDENT ASTROLOGY™, this approach is applied to the planets of the solar system (other than Earth) to see if such fundamental groupings exist. The following discussion will show that indeed, the planets come in pairs and that interestingly enough with eight planets, there are four logical pairings.

There are three immediately obvious pairings. First, Uranus and Neptune are nearly identical in size and are next to each other in orbit.

The second combination of planets—the giants Jupiter and Saturn—are also next to each other in orbit. While Saturn is smaller than Jupiter, if you consider the rings of Saturn, Saturn plus its

---

[4]Additional very small moons are continually being discovered for the four large outer planets and hence the official number of moons for these planets keeps increasing. Also, additional rings and ringlets continue to be discovered. The number of moons is based on information from the NASA Web site. The data is from late 2008.

Figure 9. Comparison of the Largest Planetary Moons.

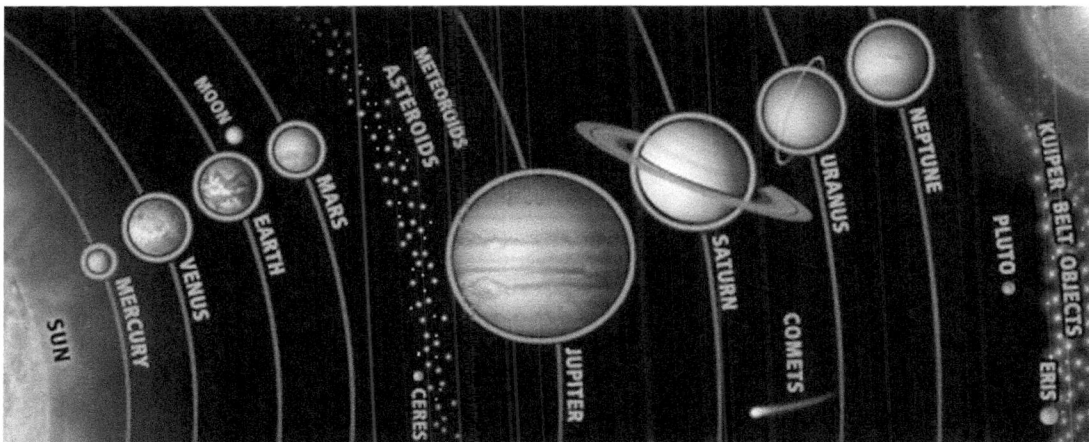

Figure 10. Planetary Orbits.

rings are actually larger in diameter than Jupiter. So depending on how you look at it, you can make the case that either is larger. Rather than push that argument, the point is that these two are more equivalent to each other from an astronomical point of view than people are generally accustomed to thinking. Also, like Uranus and Neptune, Jupiter and Saturn are next to each other in orbit.

The third pair is the inner planets of Venus and Earth. They are nearly the same in size and also next to each other in position.

Initial Review of Potential Planetary Pairs by Diameter:
1) Uranus and Neptune
2) Jupiter and Saturn
3) Venus and Earth

Lastly there is a possibility of pairing Mercury with Pluto as the two smallest, and as the planets each at the extremes of the solar system.

Additional Potential Pairings by Diameter:
a) Mercury with Mars
b) Mercury with Pluto
c) Mars with Pluto

Mercury and Pluto make a particularly good pair because they have the most elliptical and most inclined orbits in the solar system. These elliptical and inclined orbits will be explained further shortly.

So far we have looked at the most obvious characteristics of the planets' size and orbital position. Let's look at the rotational speeds and rotational tilts of their axes (such as Earth's roughly 23.5 degree tilt of its axis) to see what happens to the pairings. Here is an analysis of the speed at which the planets rotate, in other words, the length of a day on a given planet. (For detailed planetary data see Appendix I.)

1) Uranus and Neptune have nearly identical rotations of roughly two-thirds of an Earth day.
2) Jupiter and Saturn have similar rotational periods of just less than one-half an Earth day.
3) The Venus-Earth pairing does not hold well here. A Venusian day is roughly 243 sidereal days long. Not a match at all. Oddly, however, the length of a day on Earth is nearly identical to a day on Mars.
4) Mercury's day is roughly 10 times longer than Pluto's, but aside from the Venusian day that equals the Venusian year, Mercury and Pluto otherwise have the longest rotational periods.

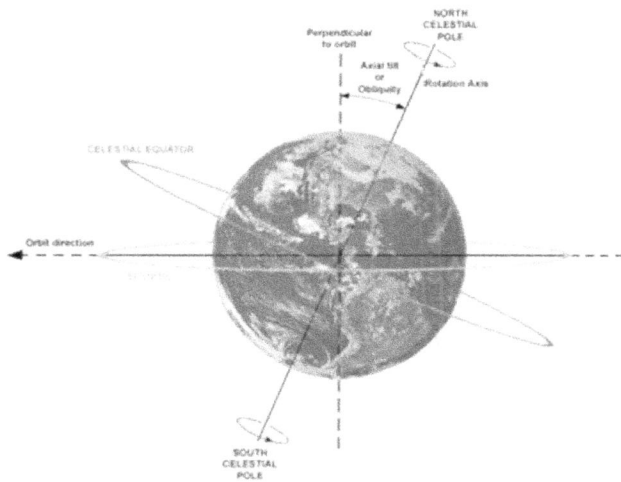

Figure 11. Axial Tilt.

Figure 11 is a diagram of the axial tilt (aka obliquity). Venus, Uranus and Pluto are essentially "upside down" relative to the tilts of all the other planets and the Sun.

The Uranus-Neptune and Jupiter-Saturn pairings are not reinforced by their axial tilt. The only planets with similar inclines are Earth, Mars, Saturn and Neptune. Of these four the only combination that we have encountered before is Earth-Mars. The Saturn-Neptune pairing is ruled out as they are not next to each other in orbit and more importantly, of very different sizes. In similar fashion neither Saturn nor Neptune would match well with Earth or Mars. Aside from these four planets with similar axial tilts, it is also true that Mercury and Jupiter have similar inclines, but since they have nothing else in common, this pairing is not pursued.

So to review, we have strong pairings between Uranus and Neptune and also between Saturn and Jupiter based on being adjacent in orbit and similar in size. These pairings are further supported by their planetary compositions, rings and moons. All four are gas giants, have many moons and many rings. By this criteria any two of the four could be matched. However Saturn is most similar to Jupiter and Neptune is most similar to Uranus. Uranus and Neptune are further paired as the first two planets not visible to the human eye[5]. Conversely Jupiter and Saturn are the last two that are visible.

Hence the Jupiter-Saturn and Uranus-Neptune pairings appear very solid, but does Earth pair with Venus or Mars and does Mercury pair with Pluto? Let's consider the Mercury-Pluto pairing first.

So far, Mercury and Pluto are the two smallest planets and the planets at the ends of the solar system. There are, however, two other physical characteristics where the pairing of Mercury and Pluto stand out: first, the orbital incline of the planets (the degree to which they orbit outside the plane of the solar system), and second, the degree of eccentricity of the orbits. The more eccentric an orbit, the less circular and more elliptical it is.

---

[5]Under truly ideal conditions, if one knows where to look and has excellent eyesight, Uranus is just barely visible to the naked eye. However, for practical purposes, it is considered the first of the "invisible" planets.

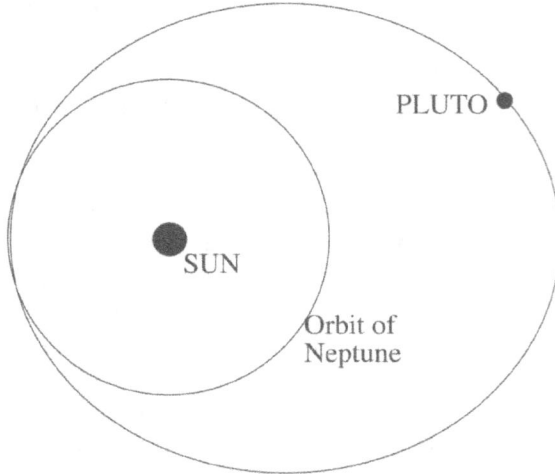

Figure 12. Pluto's Orbit.

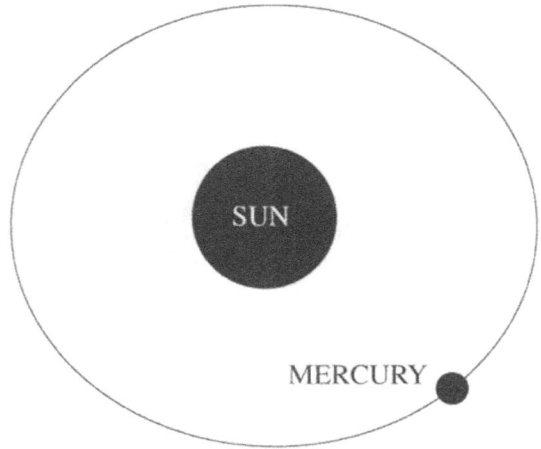

Figure 13. Mercury's Orbit.

The elliptical nature of these orbits can be seen in Figures 12 and 13. While the oval nature of Pluto's orbit is obvious, Mercury's orbit also is oval.

In terms of orbital incline (inclination from the plane of the solar system), most of the planets have 3.5 degrees inclination or less, not a great deal of variation. But the two planets with much greater orbital inclines are Mercury at 7° and Pluto at 17°.

If we look at the issue of eccentricity, it turns out that only two planets are heavily eccentric: you guessed it, Mercury and Pluto. As a side note, it is curious that Pluto is the more eccentric of the two and also is the most inclined from the plane of the solar system. Nonetheless we have answered our question: the third pairing is Mercury and Pluto. This leaves us with Venus, Earth and Mars. Since we are left with three and since Earth has things in common with both Venus and Mars, is there a pair at all among the three?

The approach of SELF-EVIDENT ASTROLOGY™ to this point has been to consider the problem from a different point of view from traditional astronomy. Up until now we have looked at the solar system objectively. What happens if we look from a subjective point of view? In the subjective point of view, Earth is the reference.

Earth as Reference

Up until the Middle Ages, philosophers of the solar system used Earth as the reference basis. It is the natural reference because it is what we understand and it is where we all are. A few hun-

12

dred years ago, astronomers found that Earth orbited the Sun and that the Sun was the center of our solar system. Earth was dropped as the reference point. Did they throw out the baby with the bath water?

While the Sun may be the center of the solar system, it is not the center of our lives; Earth is. In SELF-EVIDENT ASTROLOGY™ we return to including Earth as a very important reference point for understanding astronomy and astrology.

From Earth the most visible point in the night sky, aside from the Moon, is Venus. The most obvious point about Venus is that Venus is very nearly the same size as Earth. Since Mercury and Mars are roughly half the diameter of Earth, Venus is the only planet that might be called Earth's twin. Venus also has an orbital period which is close to an Earth year.

But we call Earth the reference point, so it can't be part of a pair observed from Earth.

Looking toward the Sun from Earth, Venus is the next closest planet. Looking away from the Sun, Mars is the closest planet to Earth. Earth is roughly fifty percent farther out from the Sun as Venus. Mars is roughly fifty percent farther out from the Sun than Earth. This similarity of orbit tends toward the idea that we should look at Venus and Mars as "equals" of Earth. Hence the idea emerges that perhaps we should pair Venus and Mars because each relates to the reference planet—Earth.

Mars is roughly half the diameter of Earth, so it is clearly not Earth's twin. But Mars has nearly the same tilt of its axis and Mars has four seasons like Earth. Mars is similar to Earth, but clearly apart from Earth as well.

Mars and Venus are the closest in size and in orbit to Earth. If Earth is the reference planet and all three of these planets represent individuals, what do we have? Venus is almost the same as Earth, or two people who are equivalent. Mars vs. Earth is two people who are different. Is this what Venus and Mars mean astrologically?

Another way to look at Venus and Mars is in respect to their atmospheres. These are the only two planets that have atmospheres that bear a resemblance to Earth's atmosphere. Venus has a very dense atmosphere. The atmosphere of Venus is so dense that it captures and retains heat from the sunlight entering the atmosphere of Venus. Thus the surface of Venus is extremely hot (as hot as Mercury's surface!)

Mars, on the other hand, may have once had an atmosphere somewhat like Earth's, but the gravity of Mars is inadequate to hold the atmosphere to the planet. Therefore, whatever atmosphere Mars may have had is now gone.

Venus holds its atmosphere and *integrates* the energy of the Sun. Only traces of the Martian atmosphere exist, so the heating of the Sun escapes (*separates*) from Mars easily. The orbit of Venus is *integrated* inside the orbit of Earth. The orbit of Mars is away from the Sun (*separated* from) relative to Earth's orbit.

Integration vs. Separation

Hence in SELF-EVIDENT ASTROLOGY™ it is suggested that Venus represents the integration of two equal individuals and that Mars represents the separation of two equal individuals. And this leads us to the pairing of Venus and Mars.

So why integration and separation? In looking for simplicity in the solar system we note that heavenly bodies are either related to one another or different from one another. It is as simple as it sounds, two heavenly bodies can be:

- identical
- similar
- different

Since no two heavenly bodies are identical to each other, we are left with similar and dissimilar. We have established that our four pairs display certain similarities. Now we examine their differences in terms of integration and separation.

The concept of pairing Venus and Mars with Earth via integration and separation from Earth suggests a very intriguing question: Do the other three pairings break down whereby one member of the pair is related to integration and the other to separation?

Jupiter and Saturn

Jupiter and Saturn are both visible in the sky. They are the most distant pair of outer planets that are clearly visible to the naked eye. What else do we know about these planets?

Let's start with the obvious. The most striking feature about Jupiter is its size, it's the biggest planet of the solar system. Jupiter has more than three times the mass of the next largest planet, Saturn.

Jupiter has the largest number of large moons. Jupiter and its moons almost appear to be a solar system unto itself. Jupiter has four inner moons similar to the "inner" planets; four large moons like the gas giants; another moon like Pluto; and a number of additional small moons similar to the Kuiper Belt. Some astronomers have suggested that if Jupiter had been larger by a factor of ten, it might have become its own sun and broken away from our solar system to form its own solar system. This is clearly suggestive of separation.

Another aspect of Jupiter that is separational in nature is that Jupiter and Mars are the planets that separate the inner planets from the outer planets with the Mars-Jupiter asteroid belt between them (separating them). (This also reaffirms Mars as a separational planet.) Okay, we see that Jupiter is separational.

Let's harken back to the idea that the inner planets have to do with individuals and the outer planets have to do with some type of "families" with all their planetary moons and rings. Let's assume we are correct that Venus and Mars have to do with individuals. What type of family/group does Jupiter represent?

Groups can be small groups like a family or large groups like a community or nation. Consider that in the night sky Jupiter and Saturn are visible and that Uranus and Neptune are not. It would seem that since family groups are far more visible to us, Jupiter and Saturn may have to do with families and therefore Uranus and Neptune may deal with communities/large groups.

If we look at Jupiter's position in the solar system, all the planets within its orbit are individual planets in nature. Moving out from the inner planets to Jupiter, it is obvious that Jupiter breaks the pattern of small planets. This break of the pattern is separational in nature. Up to this point we, tentatively have the following:

	Integration	Separation
Individual	Venus	Mars
Family	?	Jupiter

Saturn, then, is as a candidate for family/integrational. The best-known and highly visible feature of Saturn is it rings. The rings are kept in place, or so some astronomers believe, by little tiny moons that "herd" them into their position. They are integrated into the equatorial plane of Saturn. Saturn is the last visible planet. All the visible planets are inside of (integrated into) Saturn's orbit, including Jupiter, the other family planet. This suggests that families are integrated within Saturn's orbit.

If Jupiter represents the idea of a family separating (forming a new family), such as a young couple breaking away from their parents, then Saturn would seem to be integration into a family which sounds like the rearing of children by parents. Not too surprisingly these meanings are basically the same as traditional astrology, but now we have a basis in physical reality for these astrological meanings.

Uranus and Neptune

Beyond Jupiter and Saturn are Uranus and Neptune. Since Uranus and Neptune are nearly the exact same diameter and have nearly the same length of day, they are paired together.

What is the meaning of groups beyond family? As has been suggested, it would seem that we are dealing with larger groups—communities. Neptune and Uranus have families of moons like Jupiter and Saturn. Jupiter and Saturn represent single families. Because the orbits of Uranus and Neptune are outside those of Jupiter and Saturn, we have the sense of multiple families. Of course, multiple families are a community. With community being fairly certain, which planet (of Uranus and Neptune) is integrational and which would be separational?

Neptune, for a twenty-year period during every 248.5 years, is actually the farthest planet from the Sun. This was true for the period between 1979 and 1999. In this twenty-year period Pluto actually came so close to the Sun that it intercepted and went inside the mean orbit of Neptune. Because of these periods where Neptune is the planet most separated from the Sun, Neptune should be considered a potentially separational planet.

Neptune, like Mars and Jupiter, is the demarcation point for an asteroid belt. In recent years it has been found that there are a large number of fairly large asteroids beyond the orbit of Neptune. This trans-Neptunian asteroid belt is called the Kuiper Belt. So if Mars and Jupiter are related to the idea of "separation" for being adjacent to an asteroid belt, Neptune may also be considered separational (and Pluto as well for the same reason).

If "community" and "separational" are the key words for Neptune, it brings to mind subcultures like the beatniks of the 1950s, the hippies of the 1960s and the more modern subculture of computer geeks. These are communities that separate themselves from the establishment.

As Mars is the last of the individual planets (and related to the idea of separation), Neptune is also the last of the large outer gas giant planets. This similarity between Mars and Neptune again suggests the idea of separation. Like the invisibility of Neptune to the naked eye, so also small communities that break away from the establishment become relatively invisible. Often the establishment takes the point of view that these breakaway communities don't exist!

If Neptune is separational in nature, Uranus is the candidate for being integrational in nature. Uranus has two dramatic distinguishing features. One reason is that Uranus is visible (just barely) to the naked eye under favorable conditions. The second notable feature of Uranus is that it rotates at nearly a 90° angle to the plane of the solar system (actually 98°). No other planet is rotated on its axis close to 90°. This right angle feature suggests vertical structure relative to the horizontal structure of the rest of the solar system.

Community structure is fundamentally not visible. We know that a community may have a mayor and some board of elders or council members, but these are not visible like a family is. Parents are generally larger and look obviously older. Children often have a mixture of features from their parents. Visibility in families is clear; visibility in communities is not.

In the governmental structure of the USA, we have villages, towns and cities that are bound (integrated) together to form counties. Counties are grouped together to form states and the states are taken as a whole to form the country. Integration, integration, integration. In fact this is a vertical integration. Recall that Uranus spins vertically. This supports Uranus as the candidate for integrational community. Also back in the discussion of the visibility of community, we said that the mayor and the council of elders were not as visible as a family. It might be true that relative to a family structure they were less visible, but still visible, at least from time to time.

Another integrational feature of Uranus is that all the visible planets are integrated within its orbit.

Mercury and Pluto

This leaves Pluto and Mercury. If the pairings so far have the meanings of Individual, Family and Community, what is the fourth category? What category can at the same time be inside all the other planets (Mercury) and outside the other planets (Pluto)? What is the natural extension beyond community? What is both inside all communities and beyond them as well? The answer is humanity. We are all part of humanity and yet humanity is beyond any community or nation. Let's see if the fourth pairing fits this expectation.

Up to this point the pairings suggested here are not entirely new to the conventional astrologer (although the rationales are). The pairing of Mercury (left) and Pluto (right) is not generally accepted in the establishment of astrological practitioners. This pairing is now clarified.

As previously mentioned, both Pluto and Mercury have the only highly elliptical orbits of all the planets. Pluto and Mercury also have the two greatest orbital inclinations relative to plane of the solar system.

Pluto is very small, and so is Mercury. Mercury is the closest to the Sun and Pluto is the furthest. Pluto is not only invisible to the naked eye, but much more difficult to see in a telescope compared to any other planet. Mercury is also difficult to see because it is so close to the Sun. Therefore, Mercury and Pluto form a pair.

Which of the pair is integrated? If Mercury is the closest planet to the Sun, then it is the most integrated with the Sun. Pluto, not unlike Neptune, is obviously very separated from the Sun.

There is a second reason to relate the idea of integration to Mercury. Mercury has a rotation exactly two-thirds the period of one revolution around the Sun. Therefore, Mercury is tightly integrated to the Sun.

Initial Summary of Planetary Pairings

The result of our initial investigation of planetary pairings is the table below. For the first time ever the planets have been fit into a simple, logical matrix. The mysteries of astrology are beginning to evaporate.

	Integration	Separation
Individual	Venus	Mars
Family	Saturn	Jupiter
Community	Uranus	Neptune
Humanity	Mercury	Pluto

Examples of the Self-evident Planetary Meanings in Charts

Mercury in George W. Bush's chart is at 9° Leo 50′ in the first house (see Figure 14)

Mercury—Integrations into Humanity (others)—communication

First House—Individual—self

Leo—Continuation of the Individual—royal, lazy, fatherly, in prime

George W. Bush is a self-communicator. He does not need others, nor his father to sing his own praises. He does it himself and considers himself as royalty. He lets you know he thinks he is wonderful.

Venus in George W. Bush's chart is at 21° Leo 30′ in the first house

Venus—Individual Integration—beauty, partnership

First House—Individual—self

Leo—Continuation of the Individual—royal, lazy, fatherly, in prime

George W. Bush becomes himself in a partnership, especially one that allows him to appear royal and kingly.

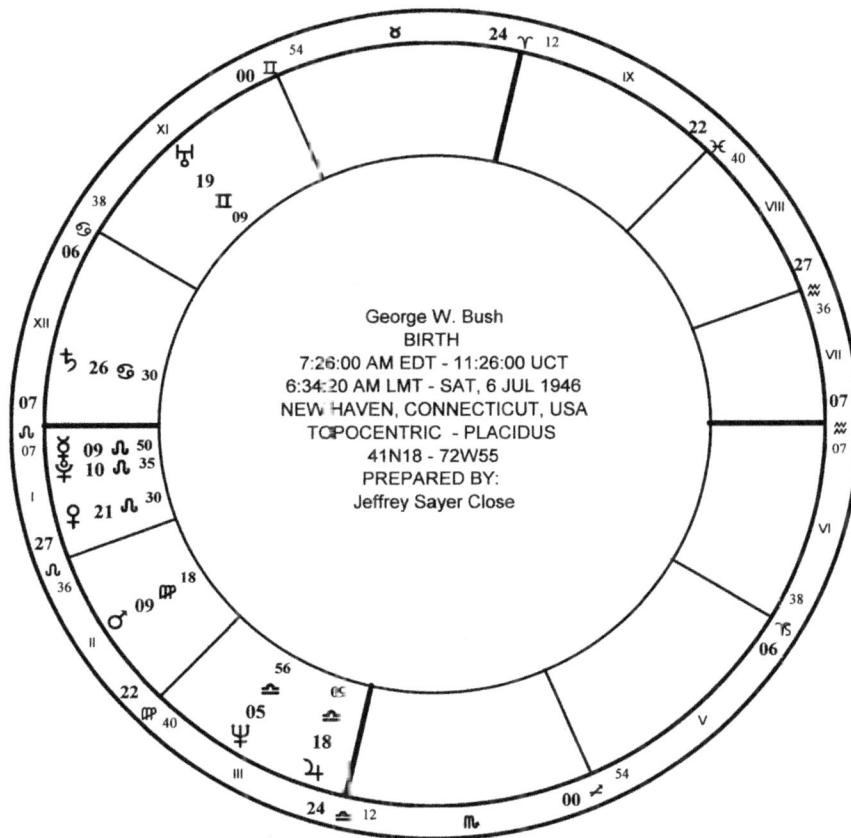

Figure 14. George W. Bush.

Mars in George W. Bush's chart is at 9° Virgo 18′ in the second house

Mars—Individual Separation—energy, war, birth

Second House—Individual Separates from Family—personal resources

Virgo—Completion of Community—hard-working perfectionist

George W. Bush, will use his own energy of his person to work hard to perfect the country. This would suggest that George W. Bush will work hard, he will throw himself into his work and seem to have tremendous energy for it. This can also mean that he is capable of sending the nation to war, essentially all by himself.

Jupiter in George W. Bush's chart is at 18° Libra 9′ in the third house

Jupiter—Family of Separations—expansion

Third House—Self is integrated into Family—learning
Libra—Start of Humanity—start of others, balance, harmony

George W. Bush integrates himself into a family (learns) by a family of separations expanding his connections to a balance of humanity (others).

Saturn in George W. Bush's chart is at 26° Cancer 30' in the twelfth house
Saturn—Integrations into Family—contraction, refinement
Twelfth House—Self separated from Community—endings
Cancer—Start of Family—protected, sensitive

George W. Bush starts making integrations (refinements) into his family by protecting those that are separated from a community.

Uranus in George W. Bush's chart is at 19° Gemini 9' in the eleventh house
Uranus—Integrations into a Community—electric, eccentric, famous
Eleventh House—Self is Integrated into Family—dreams and career
Gemini—Start of Humanity—start of others, intelligent, two-sided

George W. Bush is intelligent, but potentially duplicitous in his goal of fame.

Neptune in George W. Bush's chart is at 5° Libra 56' in the third house
Neptune—Integrations into Family—creative, confused, contrary
Third House—Self separated from Community—learning
Libra—Start of Family—harmony, balance, non-self starter

George W. Bush has creative and/or confused ways of learning balance in his life.

Pluto in George W. Bush's chart is at 10° Leo 35' in the first house
Pluto—Separate from a Universe—transformation, hidden power
First House—Self
Leo—Continuation of the Individual—royal, lazy, fatherly

George W. Bush is self-transformative and is very good at it.

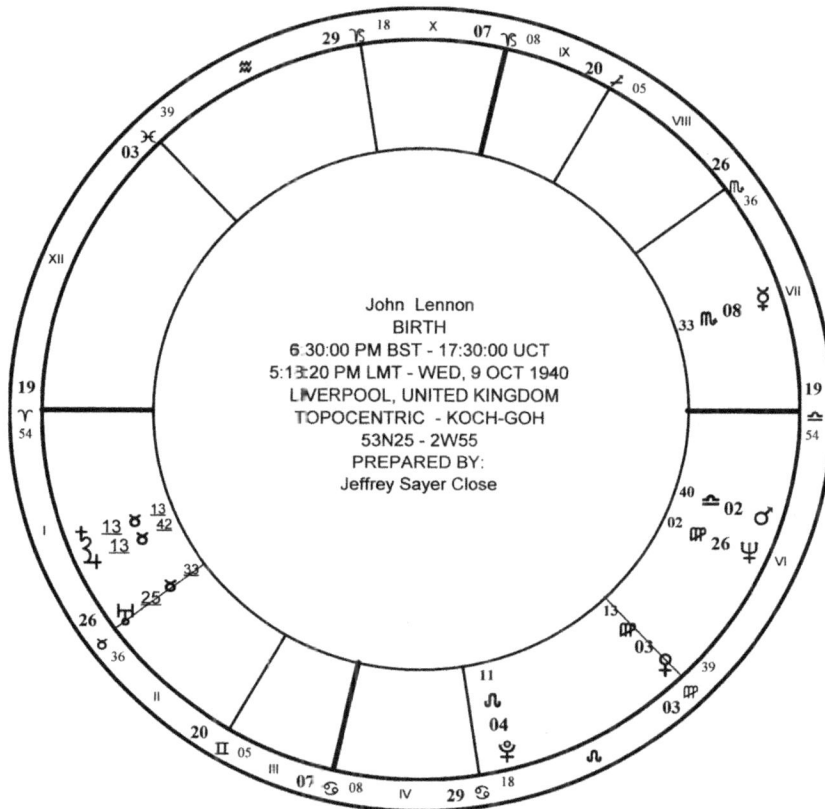

Figure 15. John Lennon

Mercury in John Lennon's chart is at 8° Scorpio 33' in the seventh house (see Figure 15).

Mercury—Integrations into Humanity (others)—communication

Seventh House—Humanity—others

Scorpio—Continuation of the Family—death, rebirth, extremes

John was extreme in his communication to humanity.

Venus in John Lennon's chart is at 3° Virgo 13' in the sixth house

Venus—Individual Integration—beauty, partnership

Sixth House—Individual Integrated into Humanity—health, perfection

Virgo—Completion of Community—hard-working perfectionist

Partnership came together for John, perfectly.

Mars in John Lennon's chart is at 2° Libra 40′ in the sixth house

Mars—Individual Separation—energy, war, birth

Sixth House—Individual Integrated into Humanity—health, perfection

Virgo—Completion of Community—hard-working perfectionist

John put all his energy into perfection of partnership and perfection by partnership.

Jupiter in John Lennon's chart is at 13° Taurus 42′ (retrograde) in the first house.

Jupiter—Family of Separations—expansion

First House—Self

Taurus—Continuation of Community—beauty, stability, value

John valued the expansion of self. This was evident in the lyrics of many of his songs. Many also believe it was John who expanded himself "beyond" the Beatles.

Saturn in John Lennon's chart is at 13° Taurus 13′ (retrograde) in the first house

Saturn—Integration into Family—contraction, refinement

First House—Self

Taurus—Continuation of Community—beauty, stability, value

John also found value in the refinement of himself.

Uranus in John Lennon's chart is at 25° Taurus 33′ (retrograde) in the first house.

Uranus—Integration into a Community—electric, eccentric, famous

First House—Self

Taurus—Continuation of Community—beauty, stability, value

John was famous for his value as an individual, his desire for the community of individuals to continue and hence a distaste for war.

Neptune in John Lennon's chart is at 26° Virgo 2′ in the sixth house.

Neptune—Integration into Family—creative, confused, contrary

Sixth House—Individual Integrated into Humanity—health, perfection

Virgo—Completion of Community—hard-working perfectionist

And John worked hard to perfect his creativity.

Pluto in John Lennon's chart is at 4° Leo 11′ in the fifth house

Pluto—Separate from a Universe—transformation, hidden power

5th House—Others Integrated into Family—friends

Leo—Continuation of the Individual—royal, lazy, fatherly

John found transformation through true friends. Friends also had hidden power over him.

Its All Just Start—Change—Stop

If you look at the Sun, Moon and the eight planets, you have the Sun starting all things, the eight planets modifying all things and then all things ending with the Moon as shown below:

Start	Change/Continue	End/Close
	Mercury	
	Venus	
	Mars	
Sun	Jupiter	Moon
	Saturn	
	Uranus	
	Neptune	
	Pluto	

Note that the Sun and Moon are in essentially circular orbits that remain constant. The eight planets move at different speeds and even appear to move backwards (retrograde) in the sky from time to time. Hence the eight planets have orbits that appear to change, but the Sun and Moon do not.

This bears repeating. If you look at the solar system from above, none of the planets ever appear to go retrograde. Retrogradation is a phenomenon that only has meaning when viewed from one of the planets. This idea is simple, the Sun and Moon don't appear to change their orbits, but all eight of the other planets appear to reverse direction periodically. Hence we have one of our simple three part fundamentals, Sun is "start," the eight planets are "change" and the Moon is "end."

Chapter Two

Symbols of the Planets Decoded

The previous chapter established the following matrix as the probable meanings of the planets:

Mercury	Humanity	Integration
Venus	Individual	Integration
Mars	Individual	Separation
Jupiter	Family	Separation
Saturn	Family	Integration
Uranus	Community	Integration
Neptune	Community	Separation
Pluto	Humanity	Separation

That this matrix is the self-evident breakdown of the meanings of the planets is further supported by the symbols that have been used by astrologers and astronomers for centuries. This chapter will show how these symbols relate directly to the keyword matrix above. What is even more fascinating is that the symbols for Mercury through Saturn come to us from antiquity. The symbols for Uranus, Neptune and Pluto were invented by different groups roughly fifty years apart from one another during the last 250 years.

There are many versions of the decoding of the meaning of the symbols of the planets. What these previous versions have in common is a lack of simplicity and lack of clarity. SELF-EVIDENT ASTROLOGY™ suggests that the true basic simplicity of the symbols can easily be seen right in the symbols themselves. Let's look at the symbols and see if obvious and consistent meanings for the symbols that are in accordance with the table above can be derived.

$$☿ ♀ ♂ ♃ ♄ ♅ ♆ ♇$$

Figure 16. The most common versions of the eight planetary symbols, left to right, Mercury, Venus, Mars, Jupiter, Saturn, Uranus, Neptune and Pluto.

The letters "PL" merged together (♇) are also used for Pluto. The fact that Pluto is the only planet with two symbols is fitting as my meaning for Pluto is being separate from all the rest. I have chosen the symbol above as the standard symbol for Pluto because Pluto's meaning is more clearly seen in this symbol. Perhaps clarity of meaning is why this is the most commonly used symbol for Pluto.

There have been many attempts to decode the symbols of the planets. Most all of these decodings are either very complicated or somewhat mystical in nature. SELF-EVIDENT ASTROLOGY™ suggests that even though these symbols were generated in different centuries by different people, even different cultures, the true basic simplicity of the symbols can easily be seen. Here are some initial observations:

1) Mercury and Pluto have the same three elements: a circle, a cross and a semi-circle
2) Jupiter and Saturn each have the same two elements, a cross and a semi-circle
3) Uranus and Neptune are both complicated and have a vertical nature.
4) Venus and Mars are centered around a large circle.

Next it is worthy of note that all eight symbols seem to break down into just four building blocks:

- circles
- semi-circles
- intersecting lines (a cross)
- arrows

Venus—Individuals Integrating

Venus (♀) is a circle on top of intersecting lines. Mars (♂) is a circle with an arrow on top of the circle. If we assume that Mars and Venus deal with individuals, then the circle would have to represent the individual.

If Venus is two individuals coming together, then the intersecting lines represent the coming together. With Mars the lines are an arrow representing separation, separation of individuals. No surprise here. These are the same meanings that we had already derived in chapter one.

Figure 17, a three-part diagram, depicts the two individuals coming together. The two individuals come from different directions: one moving vertically and one horizontally. They are the same size and therefore equals (peers) The symbol for Venus is the combination of intersecting lines representing direction and the circle for the individual. For simplicity only one circle is in the symbol. While part 3 of the diagram is the result of two individuals coming together, the diagram by itself does not show direction of travel that the intersecting lines do, so the symbol for Venus comes from part 2 of the diagram.

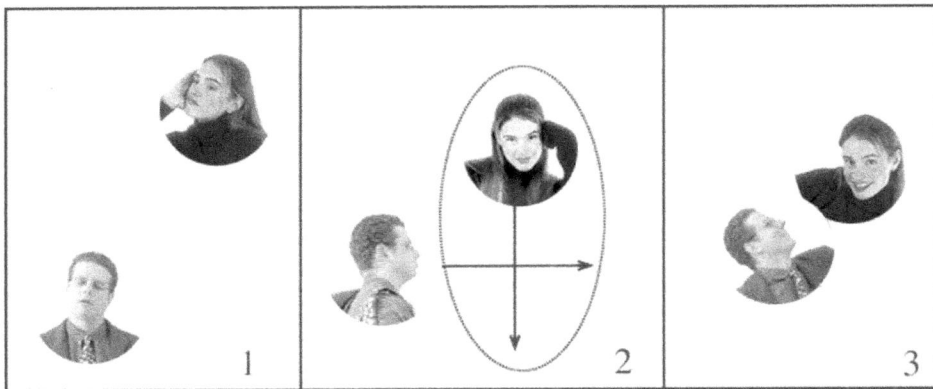

Figure 17. Individuals coming together.

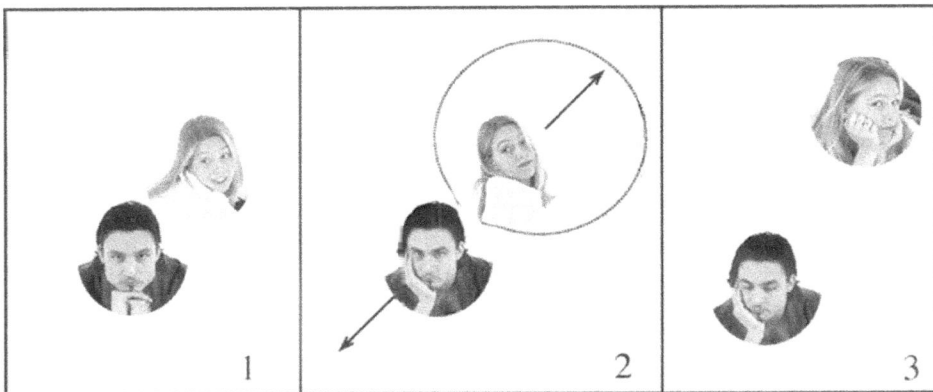

Figure 18. Individuals separating.

Mars—Individuals Separating

In Figure 18, a three-part diagram, two individuals start by being together (part 1) and then move apart in different directions (part 2). Lastly, in part 3 of the diagram they are separated. Again the symbol for Mars is taken from part 2 where the action is most evident.

Jupiter and Saturn

Moving on to Jupiter ($\mathrm{4}$) and Saturn ($\hbar$) we see the intersecting lines again. We also see the semi-circle for the first time.

Remaining consistent with Venus, the coming together of two individuals would seem to suggest the coming together of a male and female as parents.

If Saturn and Jupiter have to do with family, the intersecting lines represent the parents. Since parents don't have a full family without children, the semi-circle represents the children of a family. In Jupiter we see the family above and to the side (reminds one of the arrow of Mars that is also above and to the side) of the parents. If Jupiter is family and separation, we could be looking at the process of grown children forming new families that break away to become families on their own. This recalls the idea that if Jupiter had been a bit bigger it could have formed its own family (solar system).

Jupiter—Family Separating

In the diagram below the "children" are represented by a series of circles not unlike the many large moons of Jupiter. The children are above the parents in the diagram, reflecting the fact that

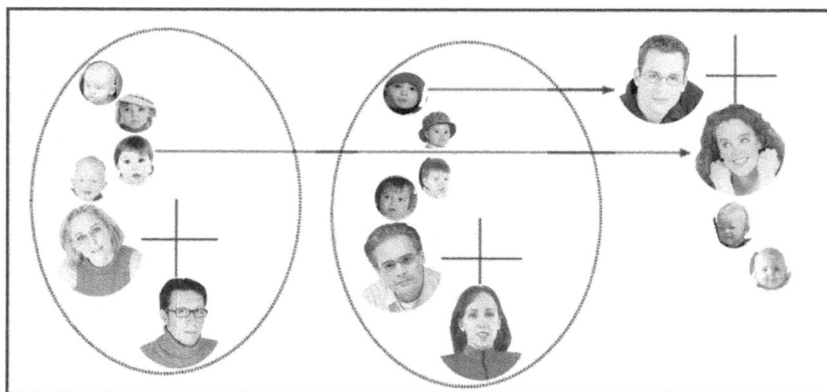

Figure 19. Family separating.

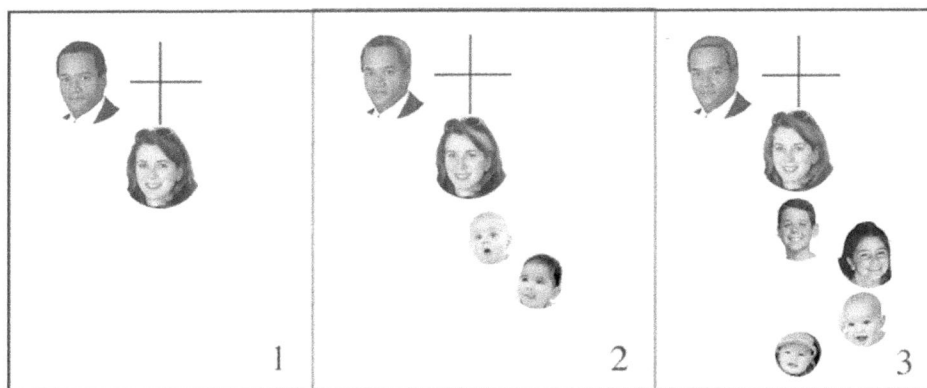

Figure 20. Family integrating.

they are moving away from the parents. As you move from left to right in the diagram below, children from each family have gone off to form a new family. Interestingly, the original families look like Jupiter's symbol, while the new family starts to look like Saturn's symbol!

Saturn—Family Integrating

Saturn is the integration of family. In Figure 20 is a set of parents starting with no children (the crossing lines), in part 2 we add two children (same as the final part of the Jupiter diagram) and then in part 3 we have four children and enough of a diagram to see the symbol for Saturn. Again, as with Jupiter, the crossing lines reflect the parents and the children form the semi-circle. Note that as with a new family the parents are superior and you might even note that with the children depicted vertically, that there is a sense of the upper (perhaps older) children being superior to the lower (younger) children.

Uranus and Neptune

Uranus (⛢), Community Integrating

Uranus is next. The left three pictures in Figure 21 show an individual not yet integrated into the community on the top. In the middle the integration has happened and in the third picture the woman is kneeling down. This gives some understanding of the feeling of Uranus. Being integrated into a community can give one the sense of oppression from which one desires to break free.

In the outline-style diagrams on the right (Figure 21), the top diagram shows an individual as the shaded circle as yet not integrated into the community. In the middle part of the diagram we see the individual integrated into the community. We also note that this integration forms a diagram (inside the "artificial" oval that has been drawn) that starts looking like the symbol for Uranus. In fact, if you were to eliminate the small circles representing the individuals and just leave the intersecting lines you have the symbol for Uranus shown to the in the bottom diagram. This bottom diagram shows the individual below the community and again the idea of wanting to break free.

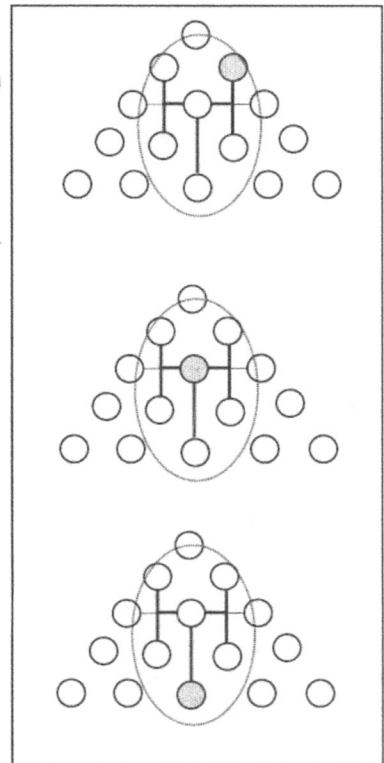

Uranus means integration into the community. The community structure, or the corporate structure for that matter, is like a pyramid. You have the mayor, or president at the top and in the next level you have, for example, the town council or advisors to the president. Then as you get lower down the pyramid, it widens out.

The idea of Uranus being divided into a lot of smaller pieces is reinforced by the largest Uranian moon. This moon appears to have been shattered and then the small pieces came together again to form the moon.

Figure 21. Community Integrating.

Neptune—Community Separating

The symbol for Neptune (Ψ) shows parents (the intersecting lines) below the semi-circle (other families) connected to three arrows all pointed away from the intersecting lines (see Figure 22). The meaning would seem to be that of a couple (or perhaps the parents representing a family) moving in a different direction from the other families.

Community Separation is the key phrase for Neptune. In the picture left of the diagram, we have replaced the semi-circle with a group of individuals that make up a community.

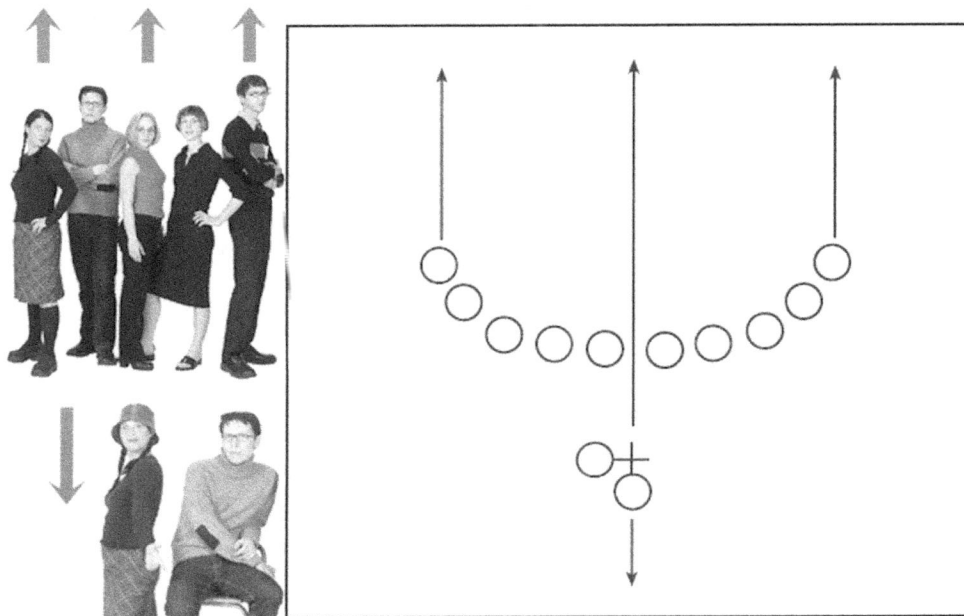

Figure 22. Community Separating.

You can see by the three upward pointing arrows that the bulk of the people are heading up, while one family is heading downward. The family heading downward is separating from the main part of the community or is becoming a sub-culture. With Neptune the concept of going against the tide is emphasized by the fact that Neptune's one very large moon is the only large moon in the solar system that orbits contrary to the rotation of the planet it orbits!

Pluto and Mercury

Pluto (Humanity Separating) is a very definitive symbol (♀). The circle (individual) is apart from family (semi-circle) and apart from individuals (cross). And since the community is built from individuals and families, if Pluto is the individual apart from both individuals and family, it is also apart from the community. In essence, Pluto seems to represent the individual apart from everything else. Mercury (☿) shows the individual (the circle) integrated between the family (semi-circle) and individuals (circle). Hence the individual is integrated into humanity.

Pluto stands for separation from humanity. Figure 23 shows an individual who, on the left, is integrated into family and community. On the right-hand part of the diagram this individual is removed from the family and community and hence is apart from humanity. Note that the right-hand diagram closely approximates the astrological symbol for Pluto. Looking to the next pair of diagrams it is interesting to note that the symbol for Mercury (☿)—made of the same

31

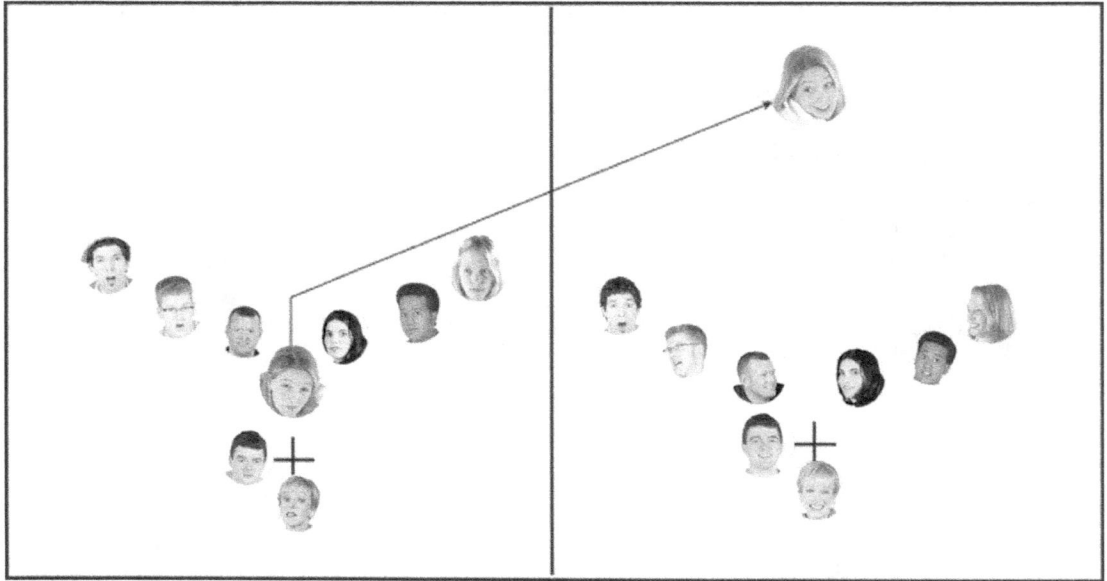

Figure 23. Separation from humanity.

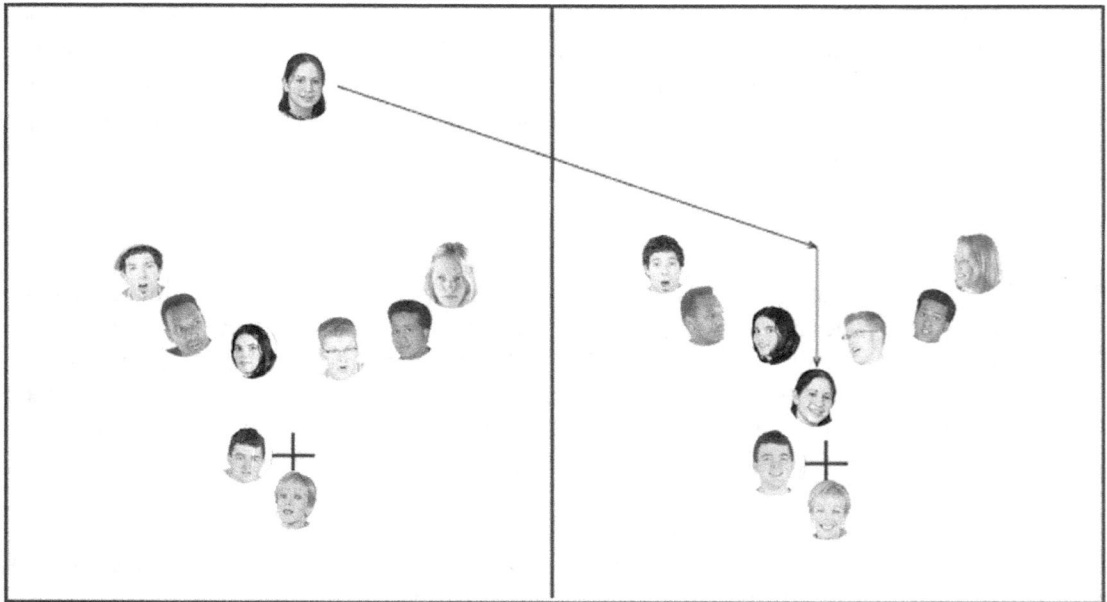

Figure 24. Integration into humanity.

components as Pluto—comes down from antiquity, but that the symbol for Pluto has only become commonly accepted recently.

With Mercury (Humanity Integrating), first is last. In Figure 24, the diagram for Mercury is simply the opposite of Pluto's. Mercury stands for the integration of all persons into humanity. In the picture above, we have replaced the semi-circle with a series of individuals. A lot of individuals make a community. As shown in the diagram, one individual gets integrated into both families and communities. Families and communities form humanity, and Mercury reflects the integration of the individual into humanity.

Lastly, Mercury has a unique characteristic that again points to the integration of all into one. Mercury is the most cratered planet. Small asteroids have been crashing (integrating) into Mercury for billions of years, leaving the face of Mercury severely pock-marked.

Rearranging our matrix with symbols, we have the following:

	Individual	Family	Community	Humanity
Integrating	♀	♄	♅	☿
Separating	♂	♃	♆	♇

To summarize the four components of the symbols, the meanings are uniformly as follows:

circle	Individual
semi-circles	Family, children
intersecting lines (a cross)	Joined individuals, parents
arrows	Direction, denoting separation

Taken both from antiquity and modern times, these symbols, it appears, are the natural symbols for the planets. This explains why the symbols completely support our matrix of planetary meanings:

	Integration	**Separation**
Individual	Venus	Mars
Family	Saturn	Jupiter
Community	Uranus	Neptune
Humanity	Mercury	Pluto

Chapter Three

The Eight Planetary Companions

The basic principle of SELF-EVIDENT ASTROLOGY™ is that the meaning of the heavens is inherent in their physical characteristics; by the same token all the bodies in the solar system have a meaning.

As can be seen in Figure 25, there are a great number of large moons in the solar system. Several are larger than Earth's Moon. Hence it is logical to ask if Earth's Moon has so much meaning, why don't astrologers have meanings for these large planetary moons?

Figure 25. Moons in the solar system.

SELF-EVIDENT ASTROLO-GY™ suggests a logical answer to this question. If we look at a planet with any rings and moons that it may have and treat this grouping as a planetary system, the meanings become more evident. Perhaps one can derive the meanings of the moons, in particular, by their characteristics. Since the vast majority of the planetary moons rotate around one of the four gas giants, there is an evident parent/child relationship between these planets and their moons. Hence it would be natural to assume a degree of inheritance of meaning from a planet to each of its moons.

The Moons of Mars

Since Mars has only two moons, Phobos and Deimos, and since Mars has to do with starting things, it is a place to start. Phobos and Deimos are both small and irregular as shown in Figure 26 and 27.

Astronomers suspect that these two moons are actually captured asteroids. Mars means the separation of two individuals. So it is appropriate that these moons probably did not come from the same gas cloud as Mars—the idea of separation is thus reinforced.

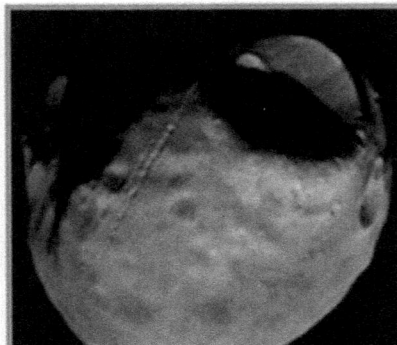

Figure 26. Phobos.

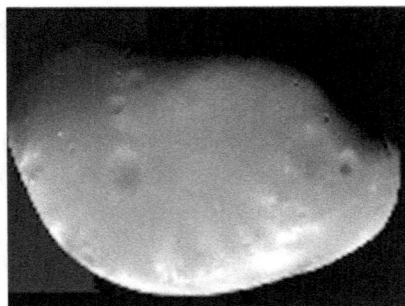

Figure 27. Deimos.

Beyond the origin of these two moons, with Phobos and Diemos we have the actual distance separating these two moons' orbits. (See Appendix II for a listing of data about the moons of our solar system.)

Phobos and Deimos are the two most separated moons (in sequential orbital positions) of all the planetary moons in all the planetary moon systems. In other words their orbits are very distant from each other.

Most importantly, the orbit of Phobos is slowly decaying. Phobos is gradually falling into Mars, but don't worry this as it won't actually happen for many millions of years. The point is that as the orbit of Phobos decays, the orbits of the two moons of Mars are separating from each other. Aside from some of the very small inner moons of Jupiter that also have decaying orbits, Triton of Neptune is the only other moon this author is aware of that is likely to be either smashed to pieces colliding with its planet, or thrown out of orbit. The emphasis in either case is separation. In SELF-EVIDENT ASTROLOGY™, Mars, Jupiter and Neptune are all related to separation.

What is most important is that Deimos and Phobos are both small, both likely captured asteroids and both separating from each other. Hence the concept of equals separating. Note we have repeated the SELF-EVIDENT ASTROLOGY™ meaning of Mars—the separation of equals.

Since the moons of Mars have the same meaning as Mars, we have a plural meaning. But is it Deimos or Phobos that has the multiple meaning of Mars? For two basic reasons, Deimos seems the proper candidate. Deimos is the furthest out (most separated from Mars and staying sepa-

Figure 28. Deimos, multiple of Mars.

rated). Phobos is decaying in orbit, this means that relative to Deimos, Phobos is separating. Note the reference point for the separation is Deimos. The symbol chosen in SELF-EVIDENT ASTROLOGY™ for Deimos, or multiple Mars, is shown to the left. The symbol is a self-evident multiple of Mars' symbol.

There are many other planetary moons in our solar system, and most are larger than Deimos and Phobos. Logic suggests that if one of the moons of Mars means multiple Mars, there are other planetary systems containing a moon that means the multiple of its respective planet.

The Moons of Jupiter

Jupiter has the distinction of being the only planetary system having four very large moons that range in size from Earth's Moon all the way to the size of Mars!

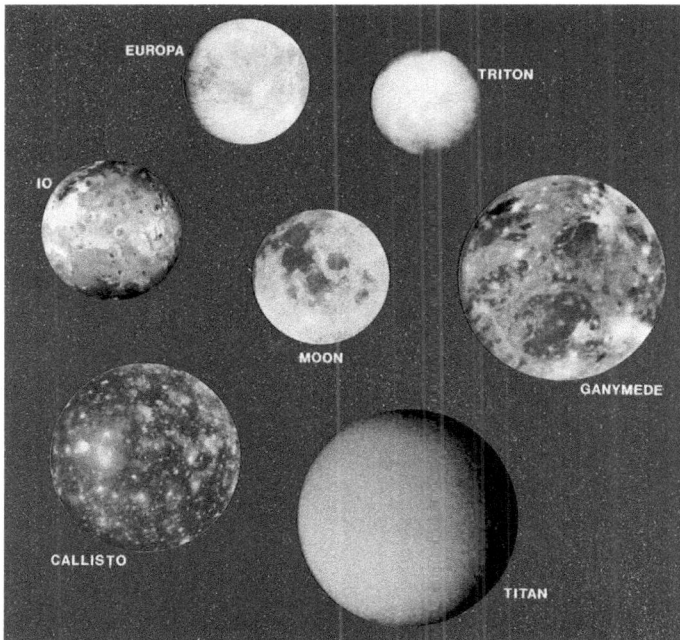

These are called the Galilean moons because they are visible via a small telescope and were discovered and recorded by Galileo about 400 years ago. The four moons are Io, Europa, Ganymede and Callisto.

These Galilean moons of Jupiter are very much a family. All four always have the same face to Jupiter, making them the only connected family of large moons in the solar system. Also, and particularly important, the orbits of the Galileans are in a synchronous arrangement.

The innermost of the four Galileans is Io. Io has an interesting relationship with the next closest moon, Europa. For every

Figure 29. The Moons of Jupiter. Shown also are Titan of Saturn and Triton of Neptune. Titan appears very large, but is slightly smaller in diameter than Ganymede of Jupiter.

two orbits of Io, Europa makes exactly one orbit around Jupiter. Io and Europa are a precise synchronous orbital arrangement.

The relationship of Io and Europa is very unusual in our solar system, but hang on, the plot thickens. As Europa goes around twice, the next more distant moon, Ganymede, goes around exactly once. So Europa and Ganymede are in precise synchronous orbits.

For purposes of understanding the next moon, Callisto, let's look at the three inner Galileans as not just being in a perfect 4:2:1 ratio; but it is also true that when Io goes around Jupiter eight times, Europa goes around four times, Ganymede goes around two times and one might expect that the last Galilean, Callisto, would go around exactly one time.

Well, yes and no. Callisto goes around almost once, but not exactly. Callisto, which has an orbital radius nearly double that of Ganymede is slightly out of this expected 8:4:2:1 synchronicity. Astronomers expect that in a few hundred million years, Callisto will join the resonance and then be locked into the perfect 8:4:2:1 arrangement.

The fact that Callisto breaks the synchronicity is striking. It puts a major emphasis on the fact that one of the family of four is separated from the others. This becomes very significant when we go back to the SELF-EVIDENT ASTROLOGY™ definition of Jupiter as "family separation." The Galileans are very true to the idea of family separation.

So how do we pick a candidate for the moon that is the "planetary companion", the multiple meaning of Jupiter. Even I at first looked to Callisto as it is the moon out of sequence. But there are more and perhaps better reasons for picking Ganymede.

Deimos was chosen as the planetary companion of Mars in part as it was the reference point from which Phobos was falling away. With the Galilieans, it is Ganymede that is the reference moon that Callisto breaks away from. The other and also important reason for picking Ganymede is that Ganymede has something important in common with Jupiter. Jupiter is the largest planet in the solar system. Ganymede is the largest moon in the solar system. Hence in SELF-EVIDENT ASTROLOGY™, Ganymede is given the honor of being called Jupiter's companion moon.

Figure 30. Ganymede.

A self-evident symbol for Ganymede is introduced with this work. Ganymede means multiple separations from a family or separation from multiple families. The symbol, shown in Figure 30, like the symbol for Deimos, is clearly built by reflecting part of the symbol of the planet, making it a multiple symbol of that planet.

It is amusing to note that Jupiter has four smaller inner moons and then four large moons (the Galileans) followed by an oddball small moon (Leda). This arrangement of the moons of Jupiter is very similar to the solar system. The four small inner moons are like the four individual inner planets inside the orbit of Jupiter. The Galileans are like the four gas Giants and Leda is small and odd, similar to Pluto. (There are many more very small moons beyond Leda—these seem similar to the Kuiper Belt, which contains many large asteroids.)

The Jovian family of moons suggests the idea that Jupiter is a reflection of our solar system as a whole. Again, as Jupiter is the largest planet, Ganymede is like Jupiter because it is the largest moon. Hence Ganymede seems to be confirmed as Jupiter's planetary companion.

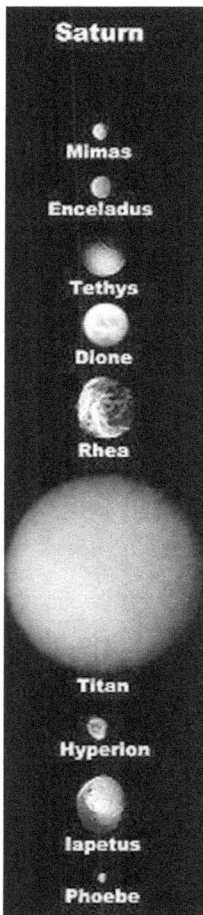

The Moons of Saturn

Saturn, like Jupiter, has a whole host of moons. Unlike Jupiter, Saturn has only one very large moon, Titan. Titan is very nearly the size of Ganymede. All the remaining moons of Saturn are much smaller than Titan. Titan is so much larger than all the others that it seems very much the dominant moon as seen in the picture of the substantive moons of Saturn in Figure 31. In fact the gravity of Titan impacts the orbits of all the moons around Saturn far more than any of them impact Titan.

In SELF-EVIDENT ASTROLOGY™, Saturn is defined as "family integration." This makes Saturn very much like the mother integrating (dominating) the children. Titan repeats this idea of domination by its sheer size. This makes Titan the clear candidate for the planetary companion of Saturn, meaning multiple family integrations or multiple integrations into a family.

Titan's similar size to Ganymede suggests another connection between Titan and Saturn. Saturn has visible rings. Looking at the solar system from the point of view of visibility, when the rings of Saturn are added to Saturn, the diameter of Saturn with its rings is greater than Jupiter. So in this one sense, Saturn appears to be the biggest planet. As Titan is clearly the largest moon of Saturn and the nearest similar in size to the largest planetary moon Ganymede, Titan is closely connected to Ganymede. The SELF-EVIDENT ASTROLOGY™ symbol for Titan is shown in Figure 32. Just as Ganymede is a "double" symbol of Jupiter, Titan's symbol is a "double" version of Saturn's symbol.

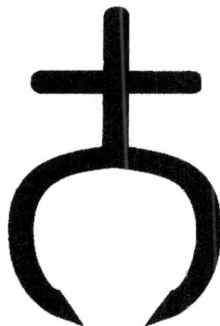

Figure 31. The Moons of Saturn.

Figure 32. Titan.

There are further situations regarding the moons of Saturn that show that Saturn has an emphasis on domination. Think of Saturn (or Titan) as a mother duckling, keeping all the children in alignment. This word alignment is closely tied to Saturn and the fits and fits the idea of family integration.

One of these alignment situations is that Janus and Epimetheus share the same orbit of 151,472 kilometers from Saturn. They are only separated by about 50 kilometers. These two satellites trade orbits about once every four years.

Another situation suggesting the idea of alignment is that three moons (Telesto, Calypso and Tethys) all share the same orbit and are separated by increments of 60°. Thus these three are in a special alignment.

A step further out away from the Sun we find the Uranian planetary system.

Moons of Uranus

The five large moons of Uranus in Figure 33 are (from left to right) Miranda, Ariel, Umbriel, Titania and Oberon. Not shown in the picture is that the moons of Uranus come in large groups (communities). Whereas Jupiter and Saturn have moons grouped in small number (families), Uranus has a group of at least ten very small inner moons, a group of five larger moons as a middle group and then a group of many more small moons beyond the larger moons pictured above. If we relate these large groups to the idea of community, then in the planetary family of Uranus, we have the concept of multiple communities. If the SELF-EVIDENT ASTROLOGY™ definition of Uranus is the "integration into a community," then which Uranian moon would be companion of Uranus, meaning the integration into multiple communities or multiple integration of communities?

The first step is to eliminate the moons of both the inner and outer groups as these moons are simply very small. This narrows our focus to the five moons shown above. These five moons break down into a group of four medium sized moons (not gigantic like Jupiter's four Galileans)

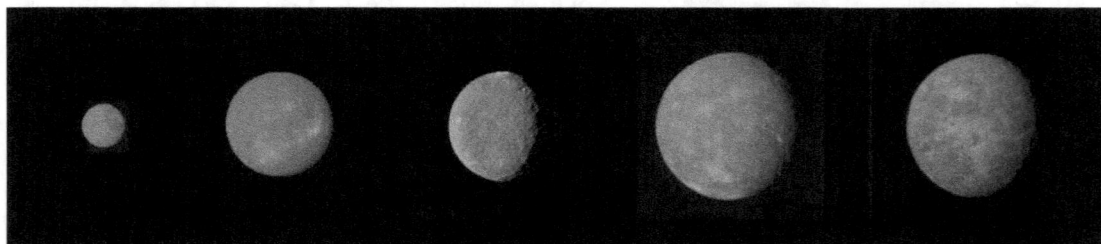

Figure 33. The Moons of Uranus.

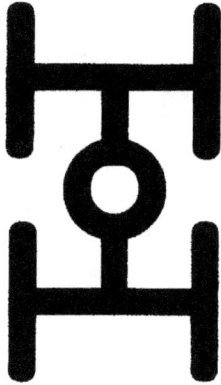

Figure 34. Miranda.

and one smaller moon—Miranda. Miranda, although smaller than the four largest moons of Uranus, is still much larger than any of the other moons of Uranus.

This odd size makes Miranda unique. Being smaller than the large Uranian moons makes it more like an individual size moon. Miranda's position puts it in a special situation. Miranda is beyond the inner groups of tiny moons but inside the orbit of the four medium sized moons. This means that Miranda is integrated between two community-sized groups of moons.

Since the SELF-EVIDENT ASTROLOGY™ definition for Uranus is "community integration" and since only Miranda fits this definition, it seems the best candidate for the planetary companion of Uranus.

The orbit of Miranda reaffirms the choice of Miranda because the four medium-sized moons beyond the orbit of Miranda (Ariel, Umbriel, Oberon and Titania) are all constantly pulling Miranda "every which way."

So when you look at Miranda as part of the community of medium-sized Uranian moons, Miranda is the one most exhibiting the concept of integration. The SELF-EVIDENT ASTROLOGY™ symbol for Miranda (Figure 34) is again a doubling of sorts of the symbol for Uranus itself.

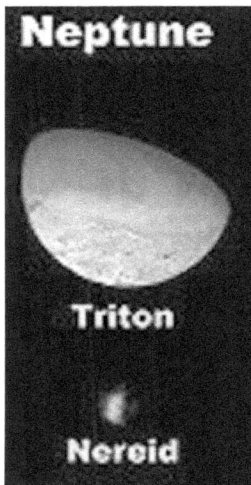

Figure 35. The Moons of Neptune.

Moons of Neptune

Moving on to Neptune, we find that Neptune's moons, like Neptune, are a bit strange and don't follow the usual patterns. Neptune has far fewer moons than the other gas giants. Neptune also has the only major moon that rotates counter to the spin of the planet it orbits.

This contrary moon is Triton. Be careful not to confuse Triton with Saturn's very large moon Titan. Except for their large size, the similarity ends with the spelling of names.

Triton may be a "captured" moon. Captured means that Triton, like the moons of Mars, probably did not form from the same gas cloud as its planet did, but was caught later in Neptune's gravitational pull when it passed very close to the planet. This would account for the contrary orbit. All the planetary moons in the solar system, except for some tiny moons, travel around their respective planet in the same direction the planet rotates. Neptune has another very strange moon, Neried, which is drastically eccentric.

Triton is not the only eccentric moon of Neptune! All of Neptune's moons have orbits averaging less than 355,000 km, but Nereid's average distance is 5,509,000 km!

The SELF-EVIDENT ASTROLOGY™ definition of Neptune is "community separation." It is a challenge to determine which is the planetary companion of Neptune, as both Nereid and Triton are different from the entire community of all planetary moons.

Figure 36. Triton.

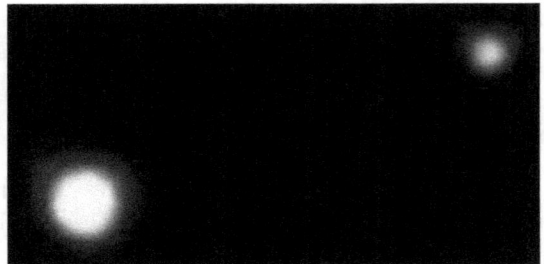

Nereid is the most eccentric, but Triton has three characteristics that seem to make it the best candidate. First Triton is much larger than Nereid or any other moon of Neptune. As we have seen with Jupiter and Saturn, the largest moon has been the companion. Triton's relationship to Neptune is very similar to Titan's relationship to Saturn, as Triton and Titan are far and away the largest moons in their planetary systems.

Second, Triton is the only major moon orbiting against the rotation of the planet and lastly, Triton is the coldest moon. Triton is colder than Neptune, Pluto or Pluto's moon Charon. This is due to the fact that Triton reflects almost all the light it receives from the Sun.

So in SELF-EVIDENT ASTROLOGY™, Triton is given the title of Neptune's Planetary Companion, meaning multiple separations from community or separation from multiple communities. The SELF-EVIDENT ASTROLOGY™ symbol (Figure 36) is a virtual "double" of the Neptune symbol.

The Moons of Pluto

Astronomers have recently discovered two very tiny moons in distant orbit around Pluto. These moons, while they reinforce the SELF-EVIDENT ASTROLOGY™ idea that Pluto means separation from the universe (of all others), each is too small and too distant to be considered as Pluto's planetary companion.

Figure 37. The Moons of Pluto.

Pluto's only other moon, Charon, is very unusual. This seems only fitting for Pluto, the unusual (even weird) planet. The fact that the International Astronomical Union kicked Pluto off the list of planets is in harmony with the SELF-EVIDENT ASTROLOGY™ meaning for Pluto, which set it apart from the universe of all others. Be Pluto a planet or not, officially, I see no change in the meaning or importance of Pluto just because a few astronomers demoted it arbitrarily.

Moving back to the question of a planetary companion, if we eliminate the two tiny distant moons of Pluto, Charon is the only choice. Does Charon deserve this title? Perhaps is does. Pluto and Charon form the only planet/moon combination

Figure 38. Pluto and Charon.

where both the planet and moon show the same face to each other at all times. This is a very tight synchronicity connecting Charon to Pluto.

Charon's size relative to its planet is, by far, the largest, Charon's diameter is a quarter of Pluto's. No other planet has a moon so relatively large. Pluto and Charon are almost a double planet. Charon seems to have no trouble fulfilling the meaning of separation from multiple universes or multiple separations from a universe.

The symbol, true to its meaning, is not a double Pluto symbol, but the Pluto symbol with Charon added to it. Charon's symbol is the only one of the planetary companions that is not a double symbol. A double symbol would have been too tall and not practical in use.

Mercury and Venus

If six planets have planetary moon companions, what of Mercury and Venus? Having covered Mars to Pluto without diverging from our pattern, we must turn to these two remaining planets. As Venus and Mercury have no known moons, where do we look for objects in the solar system that mean the multiples of each of them?

In the parlance of astronomers, planetary moons are known as "planetoids" or "minors." With Venus and Mercury we have no moons, but are there nearby minors or planetoids that might fit?

Luckily, asteroids are also included in the minors/planetoids category, and so are comets. Comets can be eliminated from consideration because of their lack of relationship to any specific planet, this leaves just asteroids.

There are two asteroid belts. One is the Mars-Jupiter Belt, the other is the Kuiper Belt. The

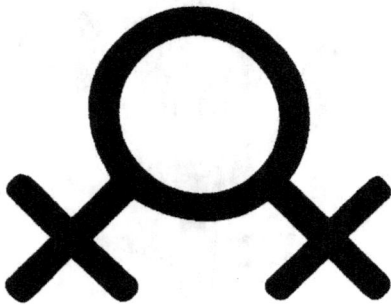

Figure 39. Juno.

Kuiper Belt is beyond Neptune and thus seems too far removed from Mercury and Venus. This leaves the Mars-Jupiter Belt.

So if we limit ourselves to the Mars-Jupiter asteroid belt, we should be looking for asteroids that have something in common with either Venus or Mercury.

Venus is the brightest planet in the night sky. One of the most highly visible asteroids, exceptionally visible for its size, is Juno. Juno was discovered around the same time as three much larger asteroids Ceres, Vesta and Pallas. Juno is half the diameter of any of the big three and there are many asteroids larger than Juno. Yet these other asteroids that are larger than Juno were not discovered for another 40 years or more. So Juno clearly stands out for brightness. If it is a match to Venus, then it should be related to multiple integrations of equals or integration of multiple equals.

Juno happens to be in orbit directly between both Ceres and Pallas. Aside from this it has a twin (in reference to its size) called Bamberga (#343). As we would expect with the companion of Venus, multiply integrated to equals, Juno is integrated to a twin and integrated to Ceres and Pallas, which are equal in visibility. The symbol for Juno looks a bit like the symbol for Venus, but it has multiple intersections. Note that this symbol is not the one generally used by astrologers for Juno. While it may take some a while to warm up to this new symbol for Juno, the older symbol does not show the connection of Juno to the other planetary companions.

The traditional symbol for Juno does resemble its meaning—multiple intersections of equals—and is thus another reason for choosing Juno as the planetary companion of Venus.

What is Mercury's Companion?

In SELF-EVIDENT ASTROLOGY™, Mercury means integration into a universe of something. Now we need to find a candidate from among the asteroids that means multiple integrations into a universe or integration into multiple universes.

My choice is Flores for two reasons. One is that of the relatively large asteroids, it is the one nearest the Sun. In other words, in the Mars-Jupiter asteroid belt, the orbit of Flores is relatively close to Mars and far from Jupiter.

The second reason, and virtually unique in its nature, is that Flores is not only an asteroid, but is an asteroid group. Flores is a fairly large asteroid surrounded tightly by a large number of tiny

Figure 40. Flores.

asteroids. It is the only large asteroid known to be in this "asteroid plus many small asteroids" combination. It is probable that Flores was once a larger asteroid that was impacted by another asteroid. Yet many of the pieces of the original Flores stayed in a tight orbit with the main asteroid, the current Flores.

Hence Flores is integrated into its own little universe of asteroids as well as being the integrated (close to the Sun) of the large asteroids in the main belt. The symbol for Flores, which is the double symbol of Mercury, shown in Figure 40, is new with SELF-EVIDENT ASTROLOGY™. It is consistent with the symbols of six of the other planetary companions and particularly similar to the new symbol for Juno. This completes the reasoning behind the eight planetary companions.

Summary of the Planetary Companions

Mercury	Flores (asteroid)	Single asteroid in many pieces
Venus	Juno (asteroid)	Brightest of the asteroids
Mars	Deimos (moon of Mars)	Reference that Phobos is separating from
Jupiter	Ganymede (moon of Jupiter)	Largest moon, "reference" next to Callisto
Saturn	Titan (moon of Saturn)	Largest moon and "mother duck"
Uranus	Miranda (moon of Uranus)	"Trapped" between two groups of moons
Neptune	Triton (moon of Neptune)	Largest moon and contrary orbit
Pluto	Charon (moon of Pluto)	Largest moon and each face each other

Planetary Companions Interpretation

The idea of planetary companions is just a nice theory unless we can put it into use. First there is a practical matter to consider. We do not have the ability to calculate the positions of the planetary moons with the same precision we expect for the planets, asteroids, and Earth's Moon. Positions of the planets have been accurately known over a period of thousands of years, but observations of the positions of the planetary moons are not as extensive. My chart calculating software, Intrepid, therefore limits their use to the period of 1850 to 2100 C.E. (A.D.)

Also, planetary moons move very fast relative to the planets. The six planetary companions that are planetary moons range in orbital periods from a bit over a day to two weeks. Deimos and Miranda are the two fastest moving. Just as the position of the Ascendant of a chart is highly influenced by knowing the right time and location for the chart, so are some of the planetary companions dependent on accurate birth times. Nonetheless, the planetary companions are still very valuable even if the position may not be known to the exact arc minute. If the time of birth is not known, only the slow moving asteroids Flores and Juno should be used.

Assuming an accurate birth time is known, how do we interpret the planetary companions? One tenet of SELF-EVIDENT ASTROLOGY™ is that the meaning of any planetary moon (aside from Earth's Moon) is partly taken from the planet's own meaning. Add to that meaning the fact that the planetary companion repeats the meaning of its planets; the meanings are multiples of the meaning of the planet in question.

So Deimos would mean multiple Mars, Ganymede multiple Jupiter and Juno would be multiple Venus. More specifically if Juno is multiple Venus and Venus means individual integration, then is Juno the integration of multiple individuals or the multiple integrations of two individuals? Since the situation lends itself to either meaning, the individual astrologer must look at both possibilities to see which fits best. Both interpretations may fit in different circumstances in a person's life.

The integration of multiple individuals is fairly clear. A baseball team would fit the concept as well as a birthday party, a platoon of infantry or just all the passengers on a bus or train. But what is the interpretation of multiple integrations of individuals? Here I would suggest you think of a series of events that have a related purpose.

A good Venus example is courtship. Relationships are built by a series of encounters. There may be a series of dates the potential couple go on. There may be a set of circumstances the two encounter that help them learn more about each other. Each of these events are part of a series that can lead to a relationship. It will generally be found that as each step along the path is taken, Juno is either aspected in the natal chart or that transiting Juno aspects a natal heavenly body. In some cases transiting Juno may be aspected by another transiting heavenly body.

Zodiacs of the Planetary Moons

I have taken a view of planetary moons that is new in its approach. While there is probably some validity to different points of view on placing planetary moons in an astrological chart, I believe the view I present is the most logical because it gives powerful and meaningful results.

A traditional view would be to look at planetary moons as bodies as seen from Earth, the same as astrologers do for the planets and asteroids that orbit the Sun. This approach has a very obvious drawback: the planetary moons always stay very close to their planets. It is rare that any planetary moon would be seen at more than a few minutes of arc separation from its parent planet.

Phobos and Deimos are so close to Mars that it is nearly impossible to measure significant distances between Mars and either of her moons. Once one looks to the moons of Uranus and beyond, the longitudinal difference between a moon and its parent is negligible.

In SELF-EVIDENT ASTROLOGY™ we have mentioned how the rings and moons of planets make them seem like their own family—their own system. Hence it makes sense to consider the longitude of a planetary moon relative to its planet. Besides, Earth's Moon is viewed that way, isn't it? The idea of viewing the planetary moons from their planets' perspectives is not my invention, but the astrological rationale is.

The result of my approach is that each planetary moon is viewed as having its own zodiac when its orbit is not close to the plane of the solar system. For example, Uranus is tilted about 98°, and its moons mostly orbit in the equatorial plane of Uranus. This means that the moons rotate vertically—perpendicular to the orbital plane—just as the Uranus rotates vertically. This restricts most of the Uranian moons to the same zodiac sign for very long periods of time from Earth's perspective and never allows them to be in more than four of the standard zodiac signs from Uranus' perspective, except for tiny increments of time.

Neptune and Pluto also have similar problems of large tilts.

Instead of using Earth's zodiac, I decided to treat each individual moon separately as astronomers do. Each moon has a first day of spring defined as its 0° Aries (the point where the moon crosses the planet's ecliptic). One might say that I have given considerable power to the children of the planets, but children are certainly known for having minds of their own!

The result of this individual treatment of planetary moons is that the moons move in a virtually constant motion through each sign of the zodiac (the zodiac as defined by that moon). This point seems self-evident: each moon has regular motion through all of the zodiac signs. If Miranda is the multiple of Uranus, how could it not be possible for it to occupy only four of the zodiac signs? If Uranus can be in any zodiac sign, then it seems logical that any of its moons could also be.

Retrograde by Inheritance

Before moving to specific examples, the SELF-EVIDENT ASTROLOGY™ approach to viewing planetary moons and whether planetary moons retrograde or not should be explained.

On the question of a planetary moon being retrograde, our own Moon is never retrograde; therefore, how could any planetary moon display retrograde motion? As viewed from their parent moons, the moons are not technically ever retrograde. However, just as it seems sensible to allow the planetary moons to be in any zodiac sign. Would it not make sense for us to think of planetary moons being retrograde?

The way I discovered to make retrogradation of the moons possible was to work again from the idea of inheritance and say that if a parent planet is retrograde, then all its moons share the qualities of the planet's apparent retrograde motion even though they are not physically retrograde. This makes some sense because when Jupiter (or any planet) goes retrograde in the sky from our perspective, the planetary moons do so at exactly the same time as their positions, in the solar system, are set by their parent planet.

Lastly, to finish the logic of this discussion, one might ask, why inheritance? Why have retrogradation of planetary moons at all. The answer is simple. We know that the planets have very different connotations in their meaning when they are retrograde. Hence it seems logical that there would also be different connotations of meanings in the planetary companions. Inheritance is the only evident way to determine when a planetary moon is considered retrograde and when it is not.

Flores and Juno are not subject to retrogradation by inheritance as they do not orbit a planet and are already seen as having periods of retrogradtion from Earth.

Planetary Companions in Action

I have included Charon in Ralph Nader's chart (Figure 41). Charon is at 26° Scorpio 13′ in the tenth house within three degrees of the Midheaven.

We know that Ralph Nader has spent much of his life discovering problems with products that consumers (in the USA) were unaware of. Many astrologers would look at Pluto in the sixth house being trine the Midheaven and suggest that this explains Mr. Nader.

It would be easy to say that Charon in the tenth house adds nothing to the position of Pluto, but it does. Pluto means separation from humanity. Mr. Nader discovered things that humanity was separate from—an understanding of all the faulty consumer products.

Charon means the multiple separations from humanity and separation from multiple large groups (humanity being a relative, not absolute, term). When looked at in this light, more of Mr. Nader can be seen. The placement of Charon in the tenth house emphasizes the fact that Mr. Nader made it his business to look for things that everyone else had missed.

Further, since Charon is related to multiple instances of things separated from humanity, or you could think of it as looking for things separated from the universe—known things of some type. Mr. Nader is well known for finding one faulty product after another. Some people would be happy to find one faulty product, but Mr. Nader made finding such faults a crusade.

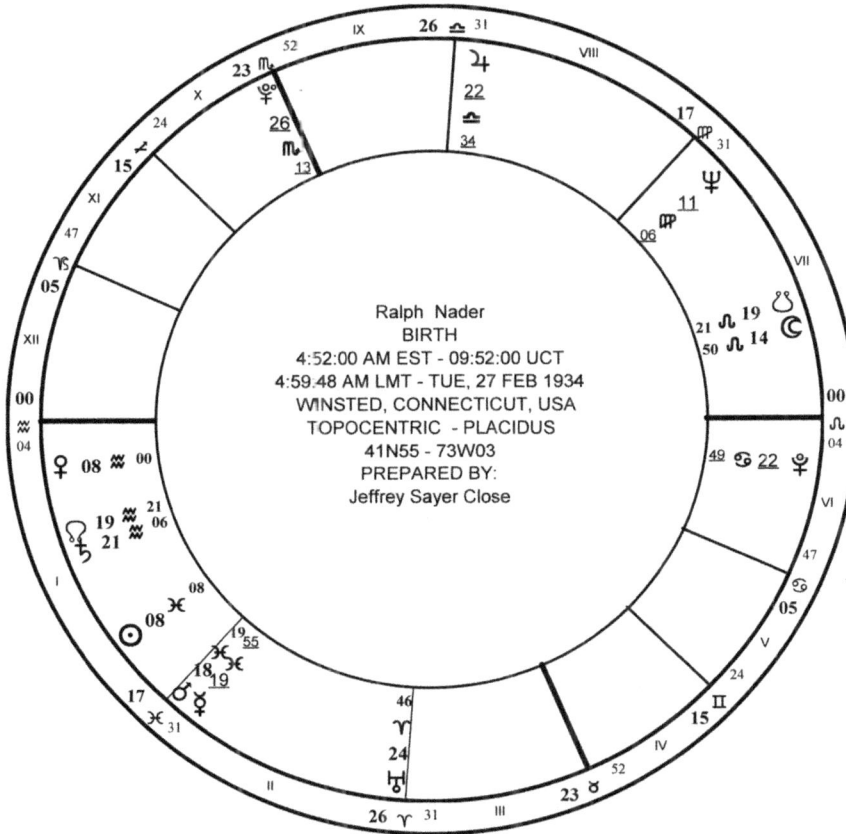

Figure 41. Ralph Nader.

Charon shows clearly that not only did Mr. Nader find those things that everyone else missed, but seemingly out of nowhere he has repeatedly run for president of the United States. So here Charon shows that Mr. Nader has multiple instances of doing something that the universe of humanity (Americans) did not expect.

Paul Newman and Deimos

In Figure 41, all eight of the planetary companions have been included, but perhaps the most striking is Deimos (multiple Mars) being within a degree of Mars. Mars is at 23° Aries 38′ and Deimos is at 22° Aries 42', both in the third house. Note that both Charon and Titan, the planetary companions of Pluto and Saturn respectively, are also in Aries in the third house.

Paul Newman is known not only for being an excellent actor, but also for at least four other major efforts in his life. Deimos is multiple cases of being separate from multiple others, or multi-

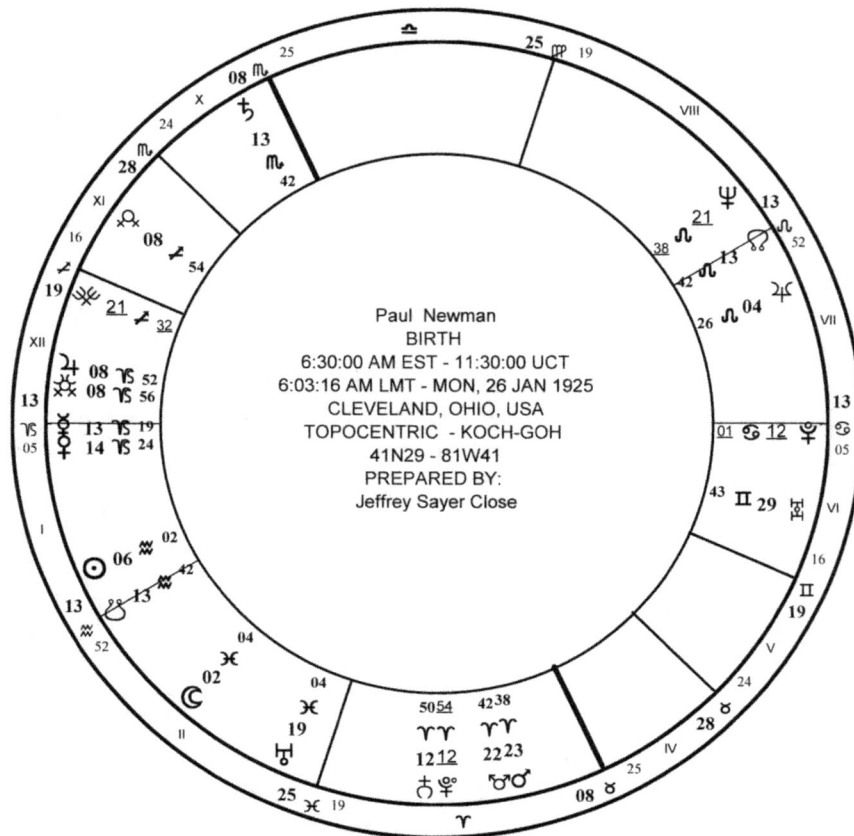

Figure 42. Paul Newman.

ple separations from another. Given that Mars and Deimos are in Aries, the desire to be an individual who starts things is very strong. Here is a list of events in Mr. Newman's life that fit better with Deimos added to Mars:

a) Excellence as an actor. A long career in acting and a reputation for being better (different from) other actors shows the repetition concept of Deimos.

b) A long and strong marriage. This separates the Newmans from most couples in the acting profession. Celebrity marriages generally don't last long, or are not strong. It is not a surprise that Newman's wife, Joanne Woodward, once described their marriage as having a large number of small arguments. This fits well with Deimos in Mr. Newman's third house (the third house being related to communication.

c) Starting a summer camp for children. Mr. Newman started the Hole in the Wall Gang camp in

Connecticut for children. This is yet another example of the multiple nature of Deimos, and work for children is fitting with Aries (individual start, which describes children).

d) Newman's Own. Most anyone in the United States has seen a food product created by the Newman's Own brand, known for the quality and healthiness of the product as well as the fact that the profits are donated to charity. Again this is an example of starting something different from what others were doing

e) Driving race cars (even at age 70). The racecar driver is great example of Deimos. He wants nothing more than to separate from other individual drivers in the race. Also, for most race car drivers, racing is a repetitive activity. Mr. Newman kept driving to age 70 and was a good racecar driver as well.

So Mars in Aries in the third house is powerful, but add Deimos and you have a person driven by the concept of competition in multiple pursuits.

It was mentioned earlier that Titan and Charon were in the same house and sign as Mars and Aries. Charon gives the flavor of being separate from a universe of others. For an actor to start a camp for children is fairly unusual. For the same person to be a top race car driver and one who starts a whole line of successful, healthy food products is, like Mr. Nader, a person who has truly done the unexpected.

Titan, the multiple of Saturn shows that Newman looked to do things that were related to integrated families. Acting integrates one into a family of other actors. Racing and competing as a food producer also do the same. And the camp for children takes children who are left out and integrates them into a family.

Summary

Planetary companions can add a great deal to a birth chart. Perhaps it offers the answer to the comment many astrologers say to me: "something is missing in the birth chart." When you add the meaning of the chart position of any planetary companion to its planet, there is so much flavoring added. It is as if you look into a field with your eyes only and see a deer. Adding the planetary companions is like looking through binoculars to see that not only is there one deer, but two baby fawns as well.

One final word on planetary companions should be noted: While we have only looked at the planetary companions in the birth chart, six of the eight planetary companions (the moons) move very quickly in a transit chart and also move relatively quickly in a day-per-year progressed chart. All six move more quickly than the Moon. But this is the subject of another work.

Chapter Four

Further Evidence and Comments
on Use in Interpretation

This chapter summarizes the study of the natural or self-evident meanings of the Sun, Moon and planets.

The questions being answered are:

- Do the planets form natural pairs?
- What do these pairs mean?
- And within each pair does one planet always relate to one concept and the other planet to the reciprocal concept?

Through the examination of these questions, we have come up with the following meanings that fulfill all three of these questions:

	Integration	**Separation**
Individual	Venus	Mars
Family	Saturn	Jupiter
Community	Uranus	Neptune
Humanity	Mercury	Pluto

Viewing the solar system in a graphical way, the meanings are as illustrated in Figure 43.

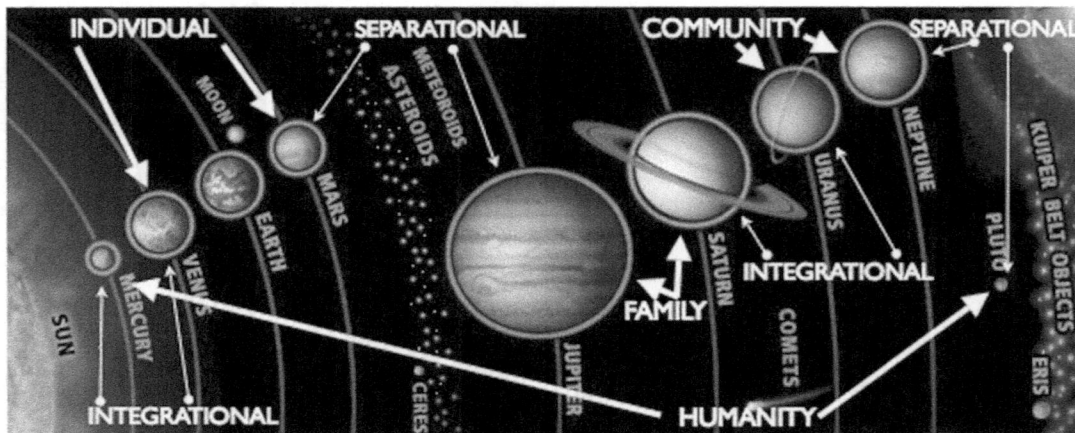

Figure 43. The solar system.

In this summary I look at each pairing and why one planet is related to the concept of integration and why the other is related to the concept of separation. In some of the summaries I show how physical characteristics, such as planetary moons reinforce the simple 2 x 4 matrix of meaning.

Further, in chapter five, I will show how this 2 x 4 matrix, along with the Sun, the Moon and the asteroid belts relate to, and in fact clarify, the relationships among the various branches of mathematics.

Neptune and Uranus (Community)
Here is a list of the reasons why these two planets should be paired together:
- They are the first two invisible planets.
- They each have several moons and rings.
- They are sequential in planetary orbit position.
- They have similar length of day.
- They are very nearly the same size.
- They generally have the same composition.

Jupiter and Saturn (Family)
Here is a list of the reasons why these two planets should be paired together:
- They are the last two visible planets.
- They each have several moons and rings.
- They are sequential in planetary orbit position.
- They have roughly similar length of day.

- They are the largest two planets and very nearly the same size (depending on how one view's Saturn's rings).
- They generally have the same composition.

Mercury and Pluto (Humanity)

Here is a list of the reasons why these two planets should be paired together:

- They are both difficult to see.
- They have the most eccentric orbits.
- They have the greatest two orbital inclines from the plane of the solar system.
- They are each at the end positions in the solar system.
- They are the two smallest planets.

Venus and Mars (Individiual)

In the case of these two planets, I have suggested that it is their relationship to Earth that makes them a pair and that Earth is the reference. So here is a look at how each of these planets compares with Earth.

First Venus:

- Venus is very similar size to Earth.
- Both have atmospheres held to a hard surfaced planet.
- They are sequential in orbit.
- They have similar compositions.
- The orbit of Venus is synchronized so when Venus makes its closest approach to Earth, the same side of Venus is always facing Earth. (The opposite is not true.)

Now Mars:

- Earth and Mars are sequential in orbit.
- They have similar compositions.
- They both have atmospheres.
- They both have four seasons (similar axial tilt).
- They both have polar ice caps.
- They both have at least one moon.

The pairings are portrayed in Figure 44

Separational vs. Integational

Now that we have reviewed the pairings, it is time to look more carefully as to why one planet of each pair is related to the concept of separation and the other related to integration.

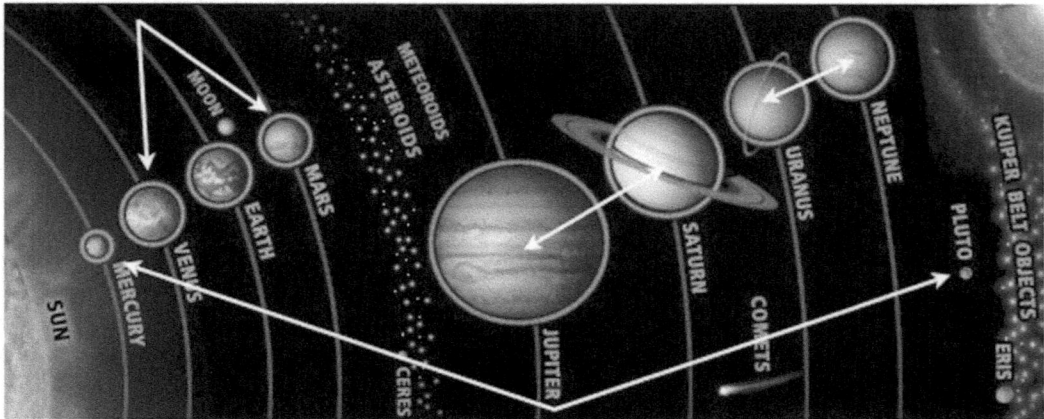

Figure 44. Pairings.

Mars as separational:

- Mars has the largest volcanoes in the solar system and volcanoes throw material out from (separate from) a planet.
- Mars has the largest known mountains in the solar system even though it is considerably smaller than Earth or Venus. Mountains, of course, suggest separation from the average surface of a planet.
- Mars is adjacent to the Mars-Jupiter asteroid belt.
- The two moons of Mars, Phobos and Deimos are separating from each other. The orbit of Phobos is slowly decaying. Astronomers expect that Phobos will either smash into Mars or be thrown out of Mars' orbit. Either event would be separational in nature. (See Figure 45.)
- Mars is the next planet beyond the reference planet (Earth) and is therefore separated from Earth. Also as viewed from Earth, Mercury and Venus are always seen near the Sun; Mars is the first planet to break that

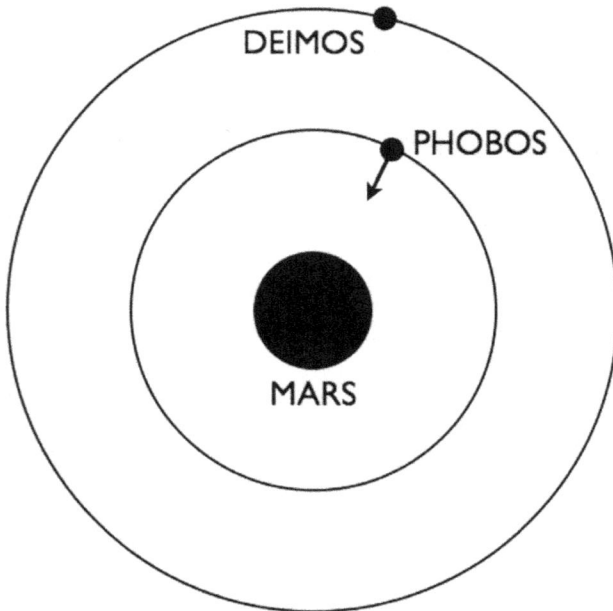

Figure 45. The Moons of Mars.

"bond" with the Sun. Unlike Mercury and Venus, Mars can occupy any position (form any angle) with respect to the Sun.

- When viewed from Earth, all eight planets appear to go backwards in the sky (called retrograde) periodically. Mars has the longest period between retrograde periods of all the planets, around twenty-six months. The next nearest, Venus, has a period between retrogrades of about eighteen months. Of course, with Venus and Mars having the longest periods between retrogrades, we have yet another reason for pairing them.
- Astronomer's theorize that Mars once had an atmosphere more like Earth's, but that with the low gravity of Mars, the atmosphere "separated" from the planet.

Venus as integrational:
- Venus has the next orbit inside of, or integrated within, Earth's orbit.
- As mentioned, Venus has an orbit that is synchronized (integrated with) Earth's
- Venus is the only planet, which has its day equal to its year. It could be said that the day and year on Venus are integrated into each other.
- The atmosphere on Venus is tightly held in, almost integrated into the surface of the planet.

Jupiter, the king of separational:

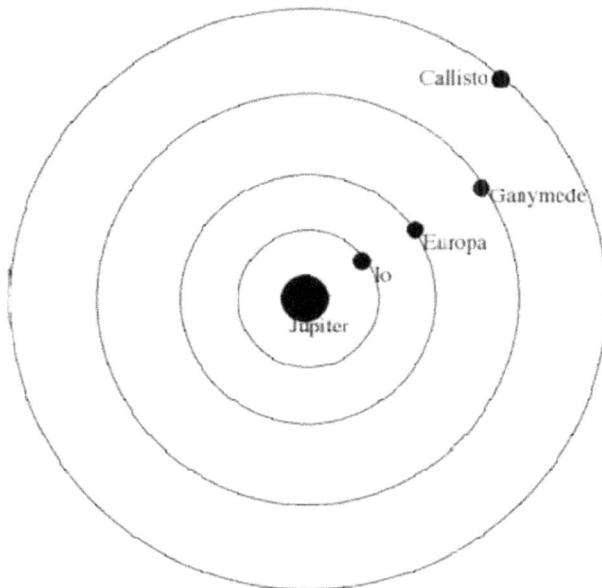

Figure 46 The Moons of Jupiter.

- Adjacent to the Mars-Jupiter asteroid belt.
- The first of the gas giant planets.
- Not counting Saturn's rings, Jupiter is the largest planet, separating it from other planets.
- Jupiter and its moons are almost like their own solar system trying to separate from our solar system.
- The four small inner moons of Jupiter, like Phobos of Mars, are slowly falling into Jupiter and will either eventually smash into Jupiter or be thrown out of Jupiter's orbit—both results being separational in nature.
- Jupiter is the only planet with more than one giant moon

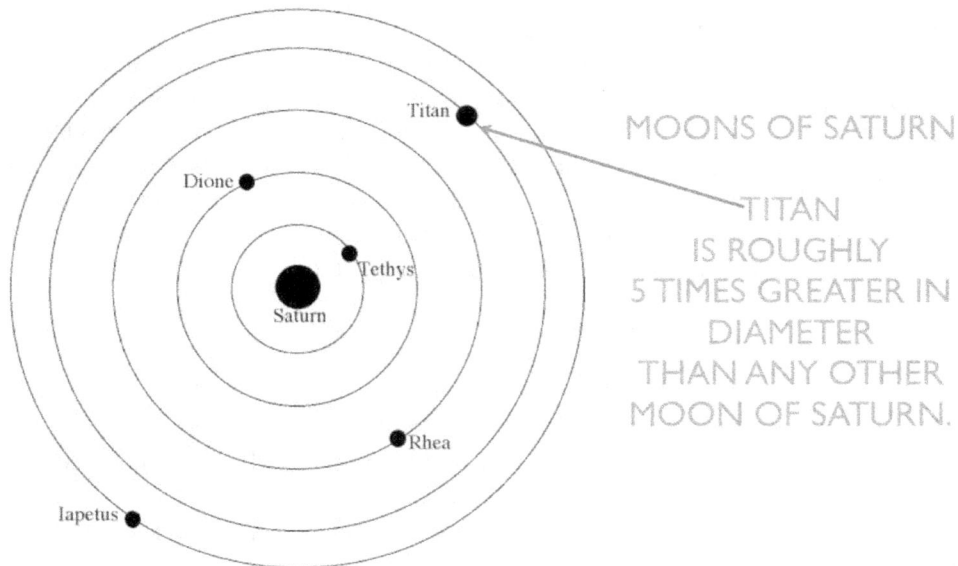

Figure 47. The Moons of Saturn.

(Jupiter has four).

- These four large moons of Jupiter strongly suggest the concept of separation. The three inner moons, Io, Europa and Ganymede are in synchronous orbits. The outermost moon, Callisto is just barely out of synchronization. Callisto emphasizes its separation from the orbital timings of the other three giant moons. (See Figure 46.)

Saturn as integrational:

- All the visible planets, save Saturn itself, are integrated within Saturn's orbit.
- Saturn has the most visible rings, which are integrated into Saturn.
- Saturn has two moons, Helene and Dione that share the same orbit (60° apart).
- Saturn also has three moons, Telesto, Tethys and Calypso that all share the same orbit, all 60° apart (Telesto leading and Calypso trailing Tethys).
- Saturn has two moons so integrated that they are constantly trading orbits (Janus and Epimetheus).
- Of all the moons of Saturn, Titan is much larger than any or the others. Titan is like a mother duck and all the other moons of Saturn are like ducklings—essentially integrated into one family as shown in Figure 47.

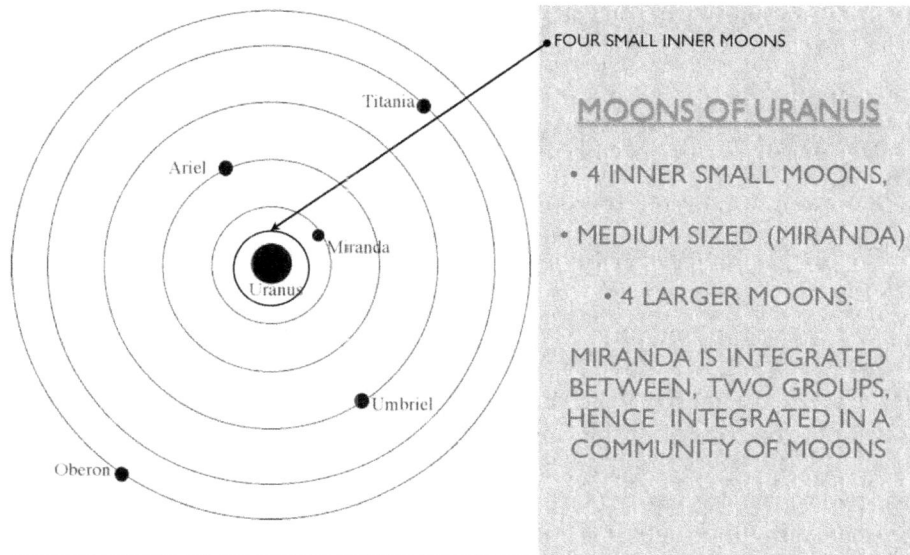

Figure 48. The Moons of Uranus.

Uranus as integrational:

- All the visible planets are integrated within the orbit of Uranus.
- Uranus is integrated into the middle of the outer planets. Jupiter and Saturn are closer to the Sun & Neptune and Pluto are further from the Sun.
- Akin to community structure, Uranus is just barely visible to the naked eye.
- Uranus rotates vertically, emphasizing hierarchy, which is a type of integration.
- The moons of Uranus also suggest the idea of integration. The moon Miranda is outside of the four small inner moons and inside the four moderately large outer moons. Miranda itself is larger than the smaller moons, but smaller than the larger moons. Hence Miranda is not a member of either group. Miranda is integrated between the two groups, lending itself to the idea of community integration. (See Figure 48.)

Neptune as separational:
- Most separated from the Sun of all the gas giants.
- For 20 of every 248.5 years Neptune is the most separated planet from the Sun.
- Neptune is adjacent to the Kuiper Asteroid Belt (Trans-Neptunian Asteroid Belt).
- The orbits of Neptune and Pluto are timed in a way that keeps them separated so that they never collide during the aforementioned twenty-year period when Pluto crosses inside the mean orbital path of Neptune. It makes sense, of course, that Pluto is also related to the idea of separational.

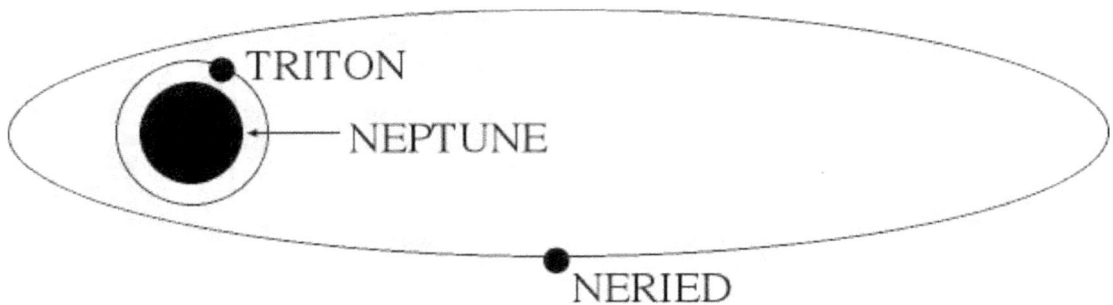

Figure 49. The Moons of Neptune.

Neptune is the reference point beyond which are communities of minor planets (comets and asteroids in the Ort Cloud and asteroids in the Kuiper Belt).

Neptune has two substantive moons that each suggest separation. Nereid has the most elliptical orbit of any planetary moon, taking it so far from Neptune that a year on Nereid is around 360 days, while a year on a typical planetary moon is around one week. The other moon, Triton, is also very strange. It is the coldest known object in the solar system and it is the only major moon that travels around its "mother" planet, contrary to the spin direction of that planet. (See Figure 49.)

Pluto as separational:
- Pluto was demoted from (separated from) the family of planets.
- For 20 of every 248.5 years Pluto is inside the orbit of Neptune.
- It is the last planet in the solar system.
- It is the smallest planet in the solar system.
- It has the most elliptical orbit in the solar system.
- It has the most inclined orbit in the solar system.
- It is the only planet discovered by an assistant astronomer.
- It is the only small planet of the outer planets.
- Pluto has a relationship with its largest moon, Charon, that is unlike any planet/moon relationship in the solar system. Charon is one-fourth the diameter of Pluto making it the largest moon relative to its planet. The two constantly show the same face to each other.

Mercury as integrational:
- Mercury is integrated within the orbits of all the other planets.
- As seen from Earth, Mercury is always within 30° of the Sun.
- Mercury's orbit and rotation are synchronized together.

More on Using Fundamental Keywords in Interpretations

When one looks at the chart below, the words are wonderful in terms of their general nature. But there needs to be comment made on how to apply these general terms to get to specific interpretations.

	Integration	**Separation**
Individual	Venus	Mars
Family	Saturn	Jupiter
Community	Uranus	Neptune
Humanity	Mercury	Pluto

One approach involves inspecting the keywords: individual, family, community and humanity. At first glance it appears that in working from individual to humanity, we just move to larger and larger groups. This is true, but there is something else we should consider.

When you look at families and communities, you are looking at vertically structured groups. Not all members of a group are equal. Families have parents above the children and communities have town elders over the citizens. Even more visible in large businesses, there is a corporate ladder. In the military community, we find the chain of command.

The difference between family and community is the size of the group. Families are small and communities are large. Note here that in interpretations, these are relative terms. Depending on the context, you could refer to the solar system as a family of planets, a community of planets, or when looked at as a whole, the nine planets are the universe of planets of our solar system.

As family and community are vertical, individual and humanity are horizontal, reflecting equality. Again like family and community, one is a small number of something (individuals) and the other is a large group of something (humanity).

This gives us some insight into the meaning of the planetary pairs:

			Size	**Orientation**
Individual	Venus	Mars	Small	Horizontal
Family	Saturn	Jupiter	Small	Vertical
Community	Uranus	Neptune	Large	Vertical
Humanity	Mercury	Pluto	Large	Horizontal

The other thing to keep in mind is that the keywords are contextual. When looking at a situation regarding humanity, keep in mind that this is more a matter of the totality of a certain type of thing. If all the people in the state of New Jersey are less than seven feet tall and one person is ten feet tall, then the ten-foot person is separate from the universe of people in New Jersey. The key is that that you have one of something that is different from a large group of equals.

The difference between family and community is a matter of families being nurturing and businesses being out in the cold world. Another aspect of family vs. community is the attitude of those involved. A tribe can act like a family but as it grows larger and larger the emphasis changes to focus on the tribal leader and the influential deputies of that leader.

It is akin to looking at feudal Europe in the Middle Ages. Families banded together to form tiny fiefdoms. The management of the fiefdom would still retain the idea of family with intermarriage between the families that were the royalty of that fiefdom. Over time the power of the king became more and more substantial. Fiefdoms merged and the small nations became communities. The communities grew into larger nations with one person at the helm, sometimes despotic, drunk with power as the community had grown so large.

The next stage in this evolution was to move to the idea of the universe of people in a nation having power. Most notably the inception of the United States fits the description of power sharing between the people and a sort of royalty. And lastly we are watching the early stages of nations being brought together by the universe of humanity that goes beyond national borders.

In using the keywords of SELF-EVIDENT ASTROLOGY™, keep in mind the context. Keep in mind whether you are dealing with a large or small group of something and that large and small are relative terms. Lastly look for vertical or horizontal organization. If you keep these concepts in mind, the key words serve well the fundamental purpose they were developed for.

Chapter Five

Mathematics, the Solar System and Rulership

If the meanings of the Sun, Moon and planets are as fundamental as I have suggested, it would not be surprising to find these fundamental patterns and meanings were found elsewhere in our lives. For something to be truly fundamental, it must be very simple. This chapter will explore the start/change/stop meanings of the solar system:

Start	Change	Stop
	Mercury	
	Venus	
	Mars	
Sun	Jupiter	Moon
	Saturn	
	Uranus	
	Neptune	
	Pluto	

and the 2 x 4 matrix of meanings:

	Integration	Separation
Individual	Venus	Mars
Family	Saturn	Jupiter
Community	Uranus	Neptune
Universe	Mercury	Pluto

have their analogies in mathematics. Further, it will be shown that SELF-EVIDENT ASTROLOGY™ leads to a standardization of the astrological rulerships (associations between the planets and the zodiac signs).

Addition and Subtraction

The most fundamental arithmetic is addition and subtraction. Addition is the bringing together (integration) of equals. In $1 + 1 = 2$ or $10 + 5 = 15$, you have the coming together (integration) of equals. The unit basis of each *individual* number is equal to any other. The integration of individuals, as has been previously shown, is related to the planet Venus.

Subtraction is the splitting apart (separation) of equals. Examples are $2 - 1 = 1$ or $15 - 5 = 10$. The separation of equals, subtraction, is related to Mars. So based on the fundamentals of SELF-EVIDENT ASTROLOGY™, it appears that Addition is related to Venus and Subtraction to Mars.

Thus the beginning of the analogy is as follows:

	Integration		**Separation**	
Individual	Venus	Addition	Mars	Subtraction
Family	Saturn		Jupiter	
Community	Uranus		Neptune	
Universe	Mercury		Pluto	

The next step in mathematics is to add and subtract larger amounts, called:

Multiplication and Division

Since Jupiter and Saturn deal with the separation and integration of families, they would seem the natural candidates to be related to these two types of mathematics.

If you add families of numbers, you are doing the same thing as multiplication. You could add $4 + 4 + 4$ and arrive at 12. This the same as adding together three families of 4's, or put in the normal form, $4 \times 3 = 12$. If Venus is the addition of equals, then SELF-EVIDENT ASTROLOGY™ would suggest that Saturn would be related to multiplication (integration of families).

If you subtract families of numbers from another number you have division. If you divide 12 by 3, the result is 4. This is the same as subtracting four sets (families) of threes from twelve. Here is what the four subtractions look like:

first, $\quad$ $12 - 3 = 9$
second, $\quad$ $9 - 3 = 6$
third, $\quad$ $6 - 3 = 3$
fourth, $\quad$ $3 - 3 = 0$

Hence the subtraction (separation) of families would relate division to Jupiter. We have all heard the concept of expanding a city by building more subdivisions. Here we have Jupiter, related to expansion alongside the word division.

To help understand how multiplication and division relate to Saturn and Jupiter respectively, here is another example. During the 1800s many wagon trains left from cities like St. Louis and headed out into the wilderness to the west. This is a good example of division. The members (generally one family per wagon) of the wagon train separated themselves (or you could say divided themselves) from what was then considered the settled portion of the United States (with apologies to Native Americans).

When looked at from the point of view of the settlers, the wagon trains amounted to the western expansion of the United States. Hence, we have the concept of division leading to the concept of expansion. The wagon trains appear to be related to the meaning of Jupiter.

The same example also helps show how multiplication relates to Saturn. Once the wagon trains start arriving at destinations such as San Francisco, the families integrated themselves into these new cities. As these families multiplied, the new cities grew. So in essence, by integration of families, we arrive at multiplication.

Words can be confusing. We refer to the growth of a city as expansion and expansion is often related to Jupiter. But note that the expansion is the result of the process of the integration of the new families.

Now our analogy is beginning to take form:

	Integration		Separation	
Individual	Venus	Addition	Mars	Subtraction
Family	Saturn	Multiplication	Jupiter	Division
Community	Uranus		Neptune	
Universe	Mercury		Pluto	

Before moving on to the higher forms of mathematics it should be noted that addition, subtraction, multiplication and division make up arithmetic. This distinction is not always something we think about, but it is true. Note that the four planets related to arithmetic are the four most visible planets. Mercury is not included because it is too close to the Sun to be easily seen.

It also makes sense that these four types of mathematics are the easiest for most people to comprehend, as they are the most visible. The forms of mathematics where you raise a number to a power or take a root of a number, like a square root, are less visible. And lastly, many people find calculus nearly incomprehensible. If you are one of these people, stay tuned, this analysis may help explain where these more complicated (less visible) branches of mathematics fit into the larger scheme of things.

Taking a Root and Raising to a Power (Exponentiation[1])

The next step is to deal with communities of numbers. The type of mathematics that deals with communities of numbers is the next natural extension from multiplication and division, those being raising a number to a power and taking the root of a number. As where multiplication is just a shorthand method for representing a great number of additions, raising to a power is a short-hand for multiplication.

For example 100^3 (one hundred cubed, not footnote #3) is 1,000,000 (one million). We could alternately describe this as 100 x 100 x 100 = 1,000,000. This is saying that the integration of communities (of 100's) is how you arrive at one million. Hence the keywords of integration and community lead us to associate Uranus here. Note, that like Uranus, the example is one that is just barely visible. By visible, in this context is meant that it is very hard for us to actually see a million of something. As an example, it you flew over a very large city, you could not actually see one million people, but you could see a great number of people and many large buildings that would suggest a population that might be one million people.

Taking the root, such as a square root, is not simple to conceptualize. Taking a root of a number can result in a type of number that we have not encountered before this point. Taking a root introduces the possibility of an irrational number.

An irrational number is one that cannot be expressed fully. Perhaps the best-known irrational number is "Pi". This is the number that starts with 3.14159... and you can keep calculating more and more decimal places and the number's "end" is never found. As best is known, you could compute Pi until the end of time and never fully describe Pi.

---

[1]Mathematical definitions from Answer.com. Two Types of exponentiation: a) Raise to a power (exponentiation). The act of raising a quantity to a power, where the power is the number of times the quantity will be multiplied by itself. b) Take a root. Square root: A number that, when multiplied by itself, will result in a given number. The square root of four is two (2 x 2); the square root of one hundred is ten (10 x 10).

Taking a root is like taking division to a larger level. The square root, for example, of one million is a thousand. Here we have divided communities (thousands being communities). Taking a root is separating from a community and hence, appears to be related to Neptune. Roots by their very nature seem almost mystical, again reminding us of Neptune.

Roots, like the creative confusion of Neptune can take some exotic forms. You can have cube roots and higher. Results of some roots give answers in the form complex numbers. A complex number is a two-part number. The first part of the number is real and the second part is imaginary. Note that imagination is a word often connected with Neptune.

Beyond the introduction of complex numbers, equations dealing with complex numbers generally result in two valid answers, not just one. One of the most famous square roots is the square root of -1 (minus one). Equations involving roots such as the square root of minus one can give these dual answer situations. As if to put icing on the cake, in the cases where there are two valid answers, one answer is real and the other is imaginary!

Let's see how our analogy to the planets has filled out:

		Integration		**Separation**
Individual	Venus	Addition	Mars	Subtraction
Family	Saturn	Multiplication	Jupiter	Division
Community	Uranus	Raise to a Power	Neptune	Take a Root
Universe	Mercury		Pluto	

Differential and Integral Calculus[2]

As can be seen in the list above we have found types of mathematics for each of the eight planets except Mercury and Pluto. The type of mathematics that may fit these last two planets is calculus. Don't panic! The fundamental idea of calculus isn't that hard.

First, calculus has two basic forms, integral and differential. Each is actually the ultimate form of addition and subtraction respectively. Just the name "integral" calculus suggests integration. The word "derivative" actually suggests, strongly in fact, the idea of separation. To derive something is to separate from what it is a part of.

---

[2]Mathematical definitions from Answer.com. Two types of calculus: a) Integral calculus. The study of *integration* and its uses, such as in finding volumes, areas, and solutions of differential equations. b) Differential calculus. The process of finding a derivative is called differentiation. The fundamental theorem of calculus states that differentiation is the reverse process to *integration*. The process of finding a derivative is called differentiation. The fundamental theorem of calculus states that differentiation is the reverse process to integration. Derivative: The limiting value of the ratio of the change in a function to the corresponding change in its independent variable.

Without trying to give an in-depth definition of integral and differential calculus, lets just say that integrals are the ultimate summation of all that something can be and that differential calculus looks for the smallest imaginable portion of something.

Here's a way to look at integral calculus. Let's say you had a swimming pool that was about thirty feet long and about twenty feet wide, but varied in depth because the bottom is curved. To find the amount of water needed to fill the pool, you could (as an approximation) think of the pool as thirty pools each one-foot long and twenty feet wide and being the depth of the pool at the middle of that one-foot slice.

This breaks the pool into simple three-dimensional segments and the volume of each portion of the pool is easy to calculate. So if you compute all thirty sections of the pool and add them together you have a reasonable approximation of the volume of water needed. Put in simple terms, integral calculus is the addition (called summation in this context) of a lot of small slices of something.

Once you integrate (add) all the slices, you have integrated the universe or totality of the pool. The terms integration and universe suggest a relation to Mercury.

The derivative is the key element of differential calculus and it is the smallest measure that something can be divided into. With a derivative you are looking at just that one small segment. If we were to look at our pool example, instead of breaking the pool into thirty slices and looking at just one of the thirty slices, the derivative would be more like breaking the pool into millions of slices and just looking at one very tiny slice.

Such a tiny slice is hard, virtually impossible, to see. Being the smallest and the hardest to see suggests Pluto. Pluto is the smallest and most invisible of the eight planets.

Here's another example: It has been said that martial arts expert Bruce Lee could punch so quickly that his punch could not be seen by a normal movie camera (thirty pictures per second). To fully capture the moment when Mr. Lee punched someone, we might need a special high-speed camera that could shoot picture frames at 300 per second. Then by careful examination we could find that one frame where the punch is landed.

This one frame when we see the desired action is a tiny segment, 1/300th of a second. This one frame out of the universe of all the frames of film taken regarding Mr. Lee's punch is like the derivative, the one tiny portion of the whole of something. You might call it the ultimate in division. Since this one frame is separated from the universe of frames of film, we have arrived at the concept of universal separation—Pluto.

So calculus isn't that hard to understand. Integral calculus is just the summation of a great number of tiny slices of something. Differential calculus is just the art of finding the exact little piece of something that you want to find.

Now the 2 x 4 matrix is complete:

		Integration			**Separation**
Individual	Venus	Addition	Mars	Subtraction
Family	Saturn	Multiplication	Jupiter	Division
Community	Uranus	Raise to a Power	Neptune	Take a Root
Universe	Mercury	Integral Calculus	Pluto	Differential Calculus

However, there is more to the solar system than the eight planets and more basic branches of mathematics than the eight we have covered.

Associations (Rulerships)

To take this analogy between the solar system and mathematics to completion, I will introduce the connection between the zodiac signs and the heavenly bodies. The full reasoning behind these specific connections are left for another book. However, the analogy of mathematics may lend itself to a portion of the reasoning behind this specific set of associations. The associations listed below that are between the heavenly bodies of the solar system and the zodiac signs are listed below. We will start with the traditional associations (aka rulerships) and work toward the SELF-EVIDENT ASTROLOGY™ version of the associations.

In traditional western astrology, such as Jim Maynard's calendar, the association between the signs and the heavenly bodies are as follows:

House	Sign	Planet
1	Aries	Mars
2	Taurus	Venus
3	Gemini	Mercury
4	Cancer	Moon
5	Leo	Sun
6	Virgo	Mercury
7	Libra	Venus
8	Scorpio	Mars and Pluto
9	Sagittarius	Jupiter

10	Capricorn	Saturn
11	Aquarius	Saturn and Uranus
12	Pisces	Neptune

There are other systems. Some astrologers, for example, do not relate Saturn and Aquarius.

In any event there are features of the list above that trouble this author. Here are some problems with these associations that our inspection of mathematics may help offer an alternative to.

a) Some planets have more than one association. Is Venus related to Taurus or Libra? Taurus and Libra are very different zodiac signs. Can one planet logically be associated with both? If so, when is any given planet related to more than one zodiacal sign? The further logical problem is this: why can't one planet be related to three or more zodiacal signs?

b) Some zodiac signs are associated with multiple planets, such as Scorpio being related to Mars and Pluto. Again Mars and Pluto are very different planets. Basically this feature has the same logical problems mentioned in item a).

c) The fact that there are multiple systems in use suggests that the fundamentals are not fully understood. If the fundamentals were understood, it would seem logical that there would only be one system—or at the very least if there were multiple systems, there would have to be a clear logic as to when to apply system #1 as opposed to when to use system #2, etc.

Here is the listing according to SELF-EVIDENT ASTROLOGY™ :

House	Sign	Planet	Branch of Mathematics
1	Aries	Mars	Subtraction
2	Taurus	Venus	Addition
3	Gemini	Mercury	Integral Calculus
4	Cancer	Moon	?
5	Leo	Sun	?
6	Virgo	Earth	?
7	Libra	Minors	?
8	Scorpio	Pluto	Differential Calculus
9	Sagittarius	Jupiter	Division
10	Capricorn	Saturn	Multiplication
11	Aquarius	Uranus	Raise to a power
12	Pisces	Neptune	Taking a Root

While the SELF-EVIDENT ASTROLOGY™ approach still needs four associations to be made, the listing above has several features that seem natural.

1) Correspondences are one to one. There are no cases of two planets for one sign or two signs for one planet.

2) All the heavenly bodies within the Mars-Jupiter asteroid belt are in the first six associations. The Mars-Jupiter belt and all planets beyond are in the second six associations.

3) Related to item 2, conventional Western astrology associates the first six houses with "I" and the second six with "we." The listing immediately above is consistent with this traditional "I/we" associations.

4) The listing above includes the Earth and the minors (a minor can be a planetary moon, an asteroid or a comet). Since the minors, as a whole, are a mass far larger than all the inner planets combined, it seems only fitting to include them. And while Earth is where we live, is there a place for a reference point in this list or is it a portion of the solar system that we are going to ignore? I don't think it can be ignored. If you look at things from Earth, you might be tempted to not include Earth. But it is equally valid to look at the solar system from outside the solar system and from that point of view Earth can not be ignored.

The next question is what are the other four types of mathematics?

Geometry and Trigonometry[3]

The easiest to start with is geometry. We shed light on an object to see what it is. The shedding of light seems like the Sun or Leo. The Sun is the center of the solar system, it is the glue that holds together the entire *geometry* of the solar system. So let's tentatively choose geometry as the branch of mathematics associated with Leo.

The definitions from Answer.com in the footnotes are perhaps inadequate for describing the difference between geometry and trigonometry. Geometry (Euclidian) is the branch of mathematics that deals with the forms of things. But what is missing from the definition is that geometry deals with how something is in its static form, not its motion.

Answer.com mentions trigonometry functions, but let's look at those functions to see how trigonometry relates to measuring forms in motion. The most basic functions of trigonometry are the

---

[3]Mathematical definitions from Answer.com. Geometry and Trigonometry. a) Geometry: The mathematics of the properties, measurement, and relationships of points, lines, angles, surfaces, and solids. b) Trigonometry: The branch of mathematics that deals with the relationships between the sides and the angles of triangles and the calculations based on them, particularly the trigonometric functions.

sine (see Figure 50) and co-sine functions. If you looked at these functions on a graph they would look a bit like ocean waves, only perfectly symmetrical. These functions describe a point on the outer edge of a wheel with the wheel turning.

You don't predict the motion of a body with geometry, but you often will do so with trigonometry. Motions are more complicated to calculate than describing the form that is in motion. The more complicated something is, the more difficult it is to understand. Things that are difficult to understand are hard to see.

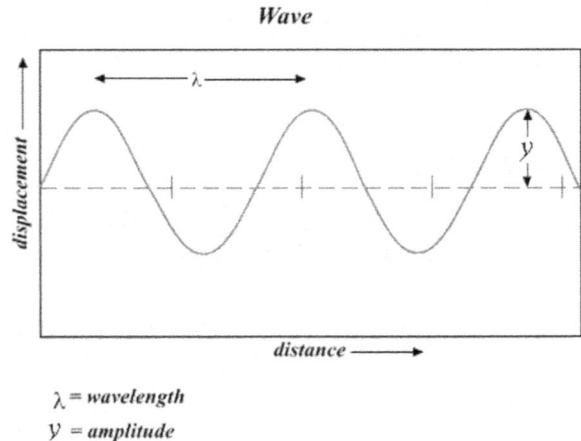

Wave

λ = *wavelength*
y = *amplitude*

Figure 50. Trigonometry.

Trigonometry is less visible than geometry, just as the Moon is less visible than the Sun. Hence the connection of the Moon and trigonometry makes sense. The connection of trigonometry to Cancer may be seen in the concept that the sign Cancer has to do with protecting and refining (like a mother rearing children). The idea of protection is a matter of hiding visibility and the refining is a matter of being able to deal with complexity.

We have now covered two of the remaining four items. The two still remaining are Virgo and Libra. The part of the chart in question is as follows:

House	Sign	Planet	Branch of Mathematics
4	Cancer	Moon	Trigonometry
5	Leo	Sun	Geometry
6	Virgo	Earth	?
7	Libra	Minors	?

Number Systems[4]

The remaining major body in the solar system is Earth. Earth is the reference point, it is where we are. It is suggested here that Earth relates to the reference point of mathematics—number systems. Systems can include polar co-ordinates as opposed to Cartesian coordinates. The most familiar number system is base 10, some computers use base 8 or base 16 and most communications systems (digital systems) use base 2.

The next question is where do we put Earth in the chart of associations? By process of elimination, our listing would suggest that Earth could be related to either Virgo or Libra. Libra does not seem to fit as it has to do with balance or the remainder. This does not suggest a reference point.

Virgo may be a different story. The time of year related to Virgo is roughly between August 21 and September 22. In the northern hemisphere, this is the time of year for the harvest. We harvest crops and this gives us the food that makes human life possible on Earth. With Earth as a reference, is not human life the underlying fundamental of all references? Virgo seems to fit with Earth. And, by extension, Earth seems related to all number systems.

Algebra[5]

The branch of mathematics relating to Libra is easy to see. It's algebra. Algebra's most basic component is the equals sign, the epitome of balance. Algebra is evaluation based on finding the remaining balance of the equation. Libra has to do with the remainder.

The scales of justice are often used as the symbol for Libra. What is it that makes the scales of justice fall one way or another? It is the remainder on one side or the other after what is equal on each side has balanced out. So if you had a balance scale and put 10 kg on each side, the scales would balance. If you added just a milligram to one side, the scales would tip. This is the idea of justice: when all else has been balanced out, what is left is what determines the ruling of a judge.

Putting Libra with the Minors is unusual in that one zodiac sign is related to multiple bodies. Even though astronomers refer to all these bodies as one class, i.e., the minors; we still have a case of one sign related to many heavenly bodies.

The idea of one to many is also one inherent in algebra. For example $y = x + 2$. We have one item on the left of the equals sign and multiple items on the right. Another example is $y = x^2 + 3z - 7$ also shows one item on the left and three items on the right.

So we have Libra related to the balance and the remainder. This makes sense in the solar system as the minors make up the balance of the bodies of the solar system.

Here is the whole chart:

---

[4]Mathematical definitions from Answer.com. Number system: Any system of naming or representing numbers, as the decimal system (base ten) or the binary system (base two). Also called *numeral system*.

[5]Mathematical definitions from Answer.com) Algebra: A branch of mathematics in which symbols, usually letters of the alphabet, represent numbers or members of a specified set and are used to represent quantities and to express general relationships that hold for all members of the set.

House	Sign	Planet	Branch of Mathematics
1	Aries	Mars	Subtraction
2	Taurus	Venus	Addition
3	Gemini	Mercury	Integral Calculus
4	Cancer	Moon	Trigonometry
5	Leo	Sun	Geometry
6	Virgo	Earth	Number Systems
7	Libra	Minors	Algebra
8	Scorpio	Pluto	Differential Calculus
9	Sagittarius	Jupiter	Division
10	Capricorn	Saturn	Multiplication
11	Aquarius	Uranus	Raise to a power
12	Pisces	Neptune	Take a Root

Now we have covered all the major branches of mathematics. The suggestion is that there may finally be a justification for a one to one relationship between the heavenly bodies and the zodiac signs by including Earth and the minors.

In turn the table above suggests that SELF-EVIDENT ASTROLOGY™ not only reveals the fundamentals underlying the traditional fundamentals, but may lend itself to the clarification of confusions in astrology that hinder the acceptance and practice of astrology.

Further Evidence of the New Rulerships/Associations

Is their more logic based on the physical characteristics of the solar system to support the new associations of planets and zodiac signs? Long accepted in traditional astrology are the four categories of zodiac signs:

Fire	Water	Earth	Air
Aries	Cancer	Capricorn	Libra
Leo	Scorpio	Taurus	Gemini
Sagittarius	Pisces	Virgo	Aquarius

These categories, also known as elements, may shed more light on the new associations.

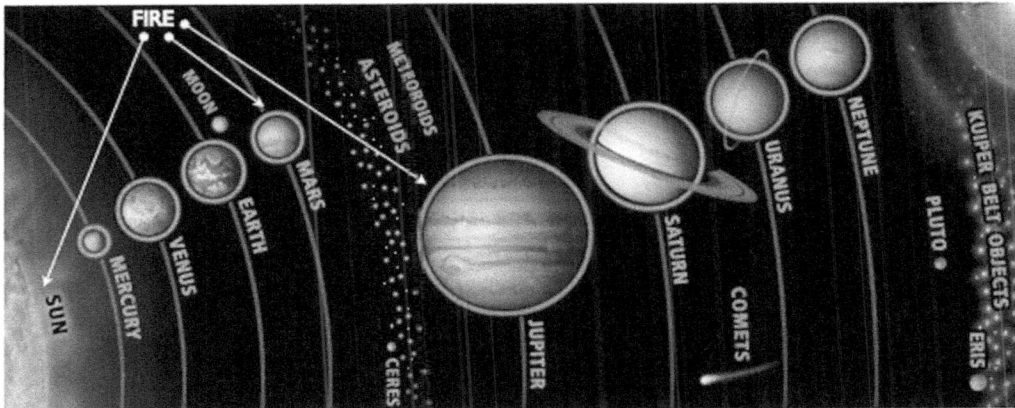

Figure 51. Fire Associations.

Figure 52. Water Associations.

Fire Associations

As shown in Figure 51, the three heavenly bodies associated with fire signs are Sun, Mars and Jupiter. What do these three have in common? Perhaps the simple answer is fire or the nature of fire.

That the Sun is full of fire is obvious. Mars is related to fire by having the largest volcanoes in the solar system. These volcanoes are particularly unusual as Mars is relatively small compared to the Earth or Venus that also have volcanoes. Jupiter is related to fire in three ways:

- Aside from the Sun it emits the most energy of any body in the solar system.
- Aside from the Sun it has the most massive magnetic energy belt.
- Jupiter has the most visible gigantic storms on its surface.

Water Associations

One of the most obvious characteristics of water comes from Earth's oceans. Whatever is in an ocean and under the surface is hard to see. So in what way are the Moon, Neptune and Pluto related to darkness? (See Figure 52.)

The simple answer for Pluto and Neptune is that they are the only two planets never visible to the naked eye. The Moon is also obviously related to darkness. We see the Moon much more clearly at night and even refer to the Moon as coming out at night.

Earth Associations

The three planets related to earth signs are Earth, Venus and Saturn. What are earth signs known for? The answer would be stability, endurance or something known for changing very slowly.

In the case of Venus, the atmosphere is exceptional in that it is held it very tightly. It is an atmosphere where almost nothing built by human hands can endure. Saturn is known for holding its rings in place. Other gas giants have rings, but Saturn's are the ones that are dramatically visible. It would seem virtually self-evident that Earth would be related to an Earth sign, it could hardly do otherwise. Besides Earth is the only planet with the earth beneath our feet that can be used to grow food. (See Figure 53.)

Air Associations

The heavenly bodies related to the air are Mercury, Uranus and the minors. The minors almost remind one of air as they have so much space between them. But there is also something all three of these bodies have in common. They are all partially visible. (See Figure 54).

Mercury is so close to the Sun it is hard to see. Uranus can be seen by the naked eye under ideal conditions only. The minors are also mixture of visible and invisible. The minors come in three flavors, each flavor has a mixed visibility:

- Comets are visible only when near the Sun.
- Planetary moons are only visible when not behind their planet or too small to be seen easily.
- Asteroids in the Mars-Jupiter belt are fairly visible. Asteroids in the Kuiper Belt are difficult to see and asteroids in the Ort Cloud are extremely hard to see.

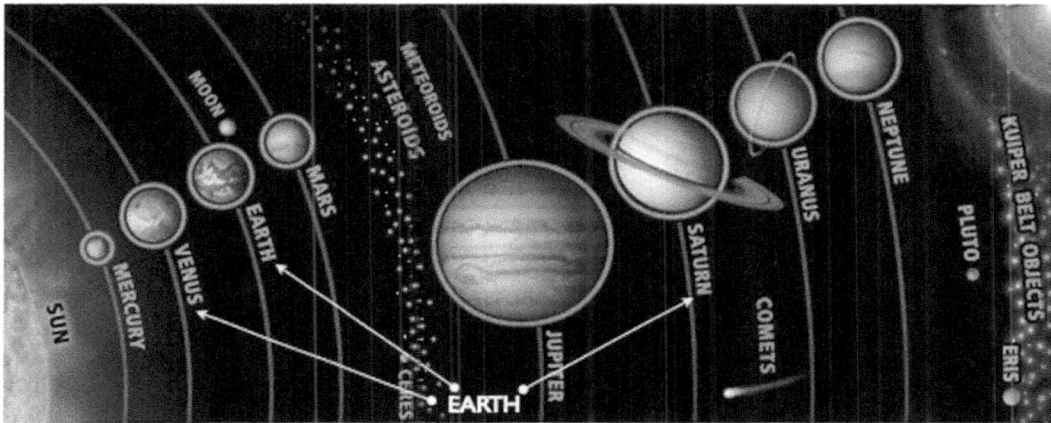

Figure 53. Earth Associations.

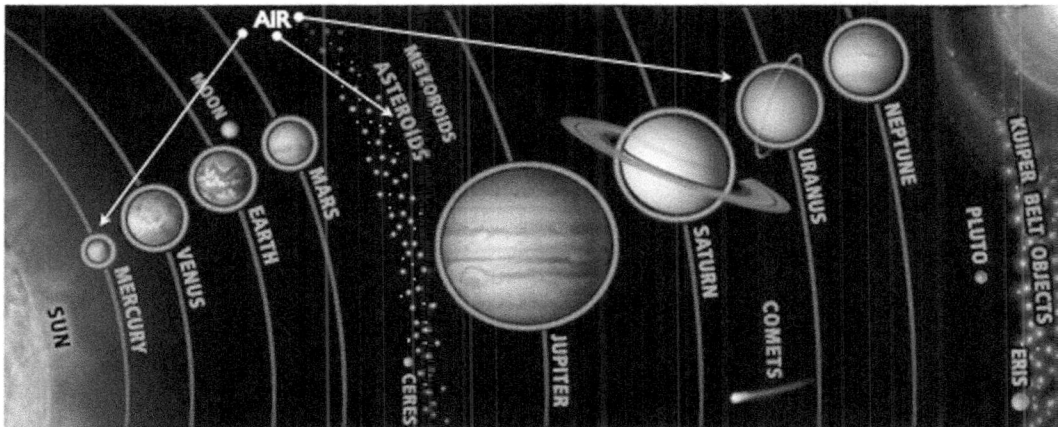

Figure 54. Air Associations.

Summary of Associations by Element

- Fire-sign related bodies are known for their energy.
- Water-sign related bodies are known for being below the surface.
- Earth-sign related bodies are known for holding things as they are.
- Air-sign related bodies are known for being a combination of visible and not visible.

Put another way, fire sign bodies are known for action as opposed to earth sign bodies known for inaction. Water sign bodies are known for being in the darkness, but the air signs are known for crossing into the light of visibility. And thus we seems to have shown the validity of the new rulerships/associations from a second point of view; making their logic very strong.

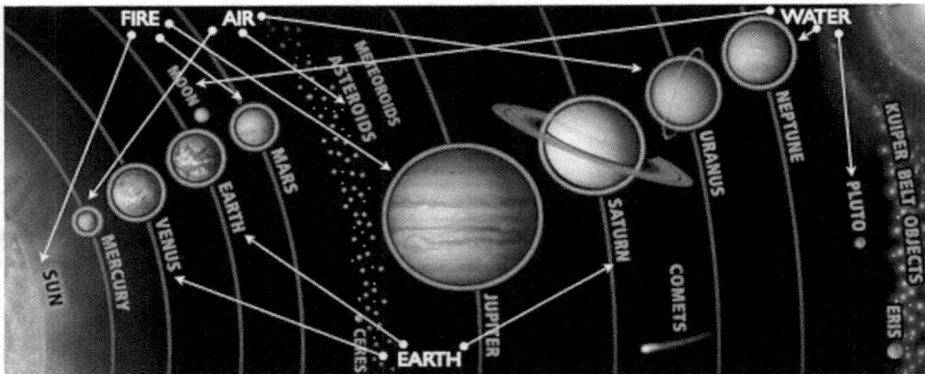

Figure 55. Summary of associations.

SELF-EVIDEDNT ASTROLOGY™ has more to reveal in terms of adjusting and adding to our traditional understanding of astrology; but these revelations are topics of other works to come.

Conclusions, Using SELF-EVIDENT ASTROLOGY™ and the Future

I believe I have demonstrated that SELF-EVIDENT ASTROLOGY™ would stand up to Thomas Jefferson's standards of what self-evident means. The physical characteristics of the planets, moons, and orbits tell us directly about their astrological meanings.

This book has looked at the heavenly bodies in the solar system (and by implication, in the larger universe also) to explain the basics of astrology and further illuminate our understanding of astrology. However, there is no claim that all of astrology can be explained by SELF-EVIDENT ASTROLOGY™.

The eight planetary companions substantively enrich astrological delineation. Each planetary companion multiplies the nature of its associated planet and indicates where you might look in a chart to see how the planet may function.

My software program, Intrepid, is the only software that properly depicts the planetary companions. Intrepid also has a natal report that includes key phrases and interpretations based on the planetary companions. At the time of this writing other interpretation reports that include SELF-EVIDENT ASTROLOGY™ are under construction.

Where do we go from here? Here is a list of topics to be covered in upcoming books on SELF-EVIDENT ASTROLOGY™:

- New fundamentals for the zodiac signs.
- Simple understanding of the meaning of the symbols of the zodiac signs.
- A new approach to house system meanings.
- The simple geometry and logic behind the nine major aspects.
- The relationship between the Western and Chinese zodiacs.
- The five new progressions of SELF-EVIDENT ASTROLOGY™.
- Using the Planetary Companions with the new progressions.
- New approaches to arc directed charts.
- Further details on using the new Rulerships listing.
- Special new techniques made easy by Intrepid Software.
- The 13 basic branches of astrology defined and explained.
- Other points in SELF-EVIDENT ASTROLOGY™ still under development.

Planet Data

This appendix reviews in greater detail how the physical characteristics of the solar system support the idea that the planets are what they mean and mean what they are. In this appendix the data comes from the NASA Web site (June 2001). The number of significant digits shown in the tables is the number of digits given by NASA.

Planetary Diameters

Planet	Radius (km)	% Difference over previous planet
Pluto	1,151	
Mercury	2,440	112 %
Mars	3,397	039 %
Venus	6,052	078 %
Earth	6,378	005 %
Neptune	24,764	289 %
Uranus	25,559	003 %
Saturn	60,268	136 %
Jupiter	71,492	019 %

Distance of Planetary Orbit from the Sun

We said that size and distance were the most obvious characteristics. Here is the data listing for distance of each planet from the Sun:

Planet	Distance (km)	% Difference over previous planet
Mercury	57,909,175	
Venus	108,208,930	087%
Earth	149,597,890	038%
Mars	227,936,640	052%
Jupiter	778,412,010	242%
Saturn	1,426,725,400	083%
Uranus	2,870,972,200	101%
Neptune	4,498,252,900	057%
Pluto	5,906,376,200	031%

In the table above the planets are listed in order going out from the Sun. The second column is the mean (average) distance of the planet from the Sun. The third column shows the relative percentage in moving from one planet to the next. For example, Venus is 87% further from the Sun than Mercury and Earth is 38% further from the Sun than Venus.

Planetary Rotation (sidereal Earth days, listed from small to large)

Planet	Day	% Difference over previous planet
Jupiter	0.41354	
Saturn	0.44401	007%
Neptune	0.67125	051%
Uranus	0.71833	007%
Earth	0.99726968	039%
Mars	1.02595675	003%
Pluto	6.38718	523%
Mercury	58.646225	818%
Venus	243.0187	314%

Rotation is the time it takes a planet to spin around once. Earth, of course, takes 24 hours for a full rotation. Jupiter and Saturn rotate at more than twice the speed of Earth, Pluto takes about a week and Mercury and Venus take most of their orbit time around the Sun for one rotation of the planet.

Orbital Incline (from ecliptic of the solar system), arranged from the smallest incline to the largest and measured in degrees.

Planet	Incline	% Difference over previous planet
Earth	0.00005	
Uranus	0.76986	1539620 %
Jupiter	1.30530	070 %
Neptune	1.76917	036 %
Mars	1.85061	005 %
Saturn	2.48446	034 %
Venus	3.39471	037 %
Mercury	7.00487	106 %
Pluto	17.14175	145 %

Orbital incline is the number of degrees different from the plane of the solar system relative to the orbit of any given planet. The most striking number is the percent change from Earth to Uranus. The large percentage shown is correct. Earth, unlike any other planet in the solar system has virtually no incline from the solar system ecliptic. This is further evidence that Earth is a reference as it has no incline from the Sun and the Sun is clearly a reference for the orbits of all planets, asteroids and comets in the solar system.

Except for Mercury and Pluto all the planets have relatively low angles of orbital incline. So here we have more evidence for matching Pluto and Mercury. Also, as always, Pluto is almost in a class by itself with a dramatic incline more than double any other planet.

Planetary Incline at Equator, listed from the smallest to the largest and in degrees.

Planet	Incline	% Difference over previous planet
Mercury	0.00	
Jupiter	3.12	(extremely large)
Earth	23.45	652 %
Mars	25.19	007 %
Saturn	26.73	006 %
Neptune	29.58	011 %
Uranus	97.86	231 %
Pluto	119.61	022 %
Venus	177.3	048 %

Planetary Incline is the tilt of the axis. This is what causes the seasons on Earth. The lack of any incline of Mercury is extraordinary and suggests the tremendous degree to which Mercury is integrated into the Sun.

Analysis here does not bear as much connection to all the data that has gone before. However, the connection between Mars and Earth is supported as they have nearly the same incline.

Jupiter and Mercury have very small inclines. Mercury has no incline, it is perfectly aligned (or integrated with) the Sun. This makes sense as Mercury is related to the concept of integration. Jupiter's connection to the Sun may be that it is the center of a "baby" solar system, making Jupiter and the Sun both centers of systems.

The equatorial incline of Saturn and Neptune are roughly similar, but the connection between these two is not clear.

Uranus is very odd in that it rotates nearly perpendicular to the plane of the solar system. This gives an emphasis to the vertical as Uranus spins perpendicular to its orbit and not more or less round and round.

Venus is also irregular in that it spins "upside down" relative to the plane of the solar system. This makes it possible for a day on Venus to equal a year on Venus (sidereally speaking).

Pluto, as usual, is not part of any group. It is neither perpendicular to nor upside down from the plane of the solar system, but simply at an odd angle different from all the other planets. This reinforces Pluto's meaning as being apart from all others.

MASS (10^{27} grams)

Planet	Weight	% Difference over previous planet
Pluto	0.013	
Mercury	0.3302	2440 %
Mars	0.64191	094 %
Venus	4.8690	659 %
Earth	5.9742	023 %
Uranus	86.849	1354 %
Neptune	102.44	018 %
Saturn	568.51	455 %
Jupiter	1,898.7	234 %

The first observation is that Jupiter is more than double the mass of all the other planets combined! This again supports the idea that Jupiter, if larger, would be on its way to being its own solar system. The idea of Venus being like the Earth is supported and the pairing of Neptune and Uranus is also supported.

While Pluto and Mercury are the least massive, it would be as easy to pair Mercury with Mars as it would be here to connect Mercury with Pluto. Naturally Pluto is in a class by itself, reinforcing this trait of Pluto. Also, and not unexpected, we see a large difference between Pluto and the inner planets vs. the outer gas giants.

Appendix II

Planetary Moon Data

Data on the followinig pages is from the NASA Web site. There are other tiny moons known to exist, not listed here.

Mars Satellite	Average Distance (000 km)	Radius (km)	Mass (kg)	Rotation	Days Orbital	Eccentric	Incline
Phobos	9,378	11	1.08e16	synchronous	0.31891023	0.015	1.0
Deimos	23,459	6	1.80e15	synchronous	1.2624407	0.0005	0.9 to 2.7

Jupiter Satellite	Average Distance (000 km)	Radius (km)	Mass (kg)	Rotation	Days Orbital	Eccentric	Incline
Metis	128	20	9.56e16		0.29478		
Adrastea	129	10	1.91e16		0.29826		
Amalthea	181	98	7.17e18	synchronous	0.49817905	0.003	00.4
Thebe	2	50	7.77e17	synchronous	0.6745	0.015	00.8
Io	422	1815	8.94e22	synchronous	1.769137786	0.004	00.04
Europa	671	1569	4.80e22	synchronous	3.551181041	0.009	00.47
Ganymede	1070	2631	1.48e23	synchronous	7.15455296	0.002	00.21
Callisto	1883	2400	1.08e23	synchronous	16.6890184	0.007	00.51
Leda	11094	8	5.68e15		238.72	0.1476	26.07
Himalia	11480	93	9.56e18		250.5662	0.1579	27.63

Lysithea	11720	18	7.77e16		259.22	0.107	29.02
Elara	11737	38	7.77e17		259.6528	0.207	24.77
Ananke	21200	15	3.82e16		631 -R	0.168	147
Carme	22600	20	9.56e16		692	0.206	164
Pasiphae	23500	25	1.91e17		735 -R	0.378	145
Sinope	23700	18	7.77e16		758 -R	0.275	153

Saturn Satellite	Average Distance (000 km)	Radius (km)	Mass (kg)	Rotation	Days Orbital	Eccentric	Incline
Pan	134	10	?		0.5750		
Atlas	138	14	?		0.6019	0.000	0.3
Prometheus	139	46	2.70e17		0.6130	0.003	0.0
Pandora	142	46	2.20e17		0.6285	0.004	0.0
Epimetheus	151	57	5.60e17	synchronous	0.6942	0.009	0.34
Janus	151	89	2.01e18	synchronous	0.6945	0.007	0.14
Mimas	186	196	3.80e19	synchronous	0.94242181	0.0202	1.53
Enceladus	238	260	8.40e19	synchronous	1.37021785	0.00452	0.00
Tethys	295	530	7.55e20	synchronous	1.88780216	0.0000	1.86
Telesto	295	15	?		1.8878 leads Tethys		
Calypso	295	13	?		1.8878 trails Tethys		
Dione	377	560	1.05e21		2.73691474	0.00223	0.02
Helene	377	16	?		2.7369	0.005	0.0
Rhea	527	765	2.49e21	synchronous	4.51750043	0.00100	0.35
Titan	1222	2575	1.35e23	synchronous	15.9454206	0.02919	0.33
Hyperion	1481	143	1.77e19		21.2766088	0.104	0.43
Iapetus	3561	730	1.88e21	synchronous	79.3301825	0.02828	14.72
Phoebe	12952	110	4.00e18		550.48 -R	0.16326	177°

Janus and Epimetheus share the same orbit. They are only separated by about 50 kilometers (31 miles). As these two satellites approach each other they exchange momentum and trade orbits; the inner satellite becomes the outer and the outer moves to the inner position. This exchange happens about once every four years. Janus and Epimetheus may have formed from a disruption of a single parent to form co-orbital satellites.

Telesto and Calypso are called the Tethys Trojans because they circle Saturn in the same orbit as Tethys, about 60 degrees ahead of and behind that body. Telesto is the leading Trojan and Calypso is the trailing Trojan.

Dione has probably been tidally locked in its current position for the past several billion years.

Helene and Dionne are in same orbit 60° apart. Hyperion is the most irregular large moon probably had previous major collision. Its rotational period varies. Phoebe is probably a captured asteroid.

Uranus Satellite	Average Distance (000 km)	Radius (km)	Mass (kg)	Rotation	Days Orbital	Eccentric	Incline
Cordelia	50	13	?		0.3350338	0.00026	0.08
Ophelia	54	16	?		0.376400	0.0099	0.10
Bianca	59	22	?		0.43457899	0.0009	0.19
Cressida	62	33	?		0.463570	0.000	0.04
Desdemona	63	29	?		0.47364960	0.00013	0.11
Juliet	64	42	?		0.49306549	0.00066	0.07
Portia	66	55	?		0.51319592	0.0000	0.06
Rosalind	70	27	?		0.55845953	0.0001	0.28
Belinda	75	34	?		0.62352747	0.0007	0.03
Puck	86	77	?		0.76183287	0.00012	0.32
Miranda	130	236	6.30e19	synchronous	1.41347925	0.0027	4.2
Ariel	191	579	1.27e21	synchronous	2.52037935	0.0034	0.3
Umbriel	266	585	1.27e21	synchronous	4.1441772	0.0050	0.36
Titania	436	789	3.49e21	synchronous	8.7058717	0.0022	0.14
Oberon	583	761	3.03e21	synchronous	13.4632389	0.0008	0.10
Caliban	7164	40	?		579.379	0.082	139.2
Sycorax	12174	80	?		1284	0.509	152.7
Stephano	7948	15	?				
Setebos	17681	20	?				
Prospero	16568	20	?				

Cordelia, Ophelia are shepards. Ariel is the brightest. Umbriel is the darkest.

Miranda—Its surface is unlike anything in the solar system with features that are jumbled together in a haphazard fashion.

Neptune Satellite	Average Distance (000 km)	Radius (km)	Mass (kg)	Rotation	Orbital	Eccentric	Days Incline
Naiad	48	29	?		0.294396	0.000	4.74
Thalassa	50	40	?		0.311485	0.000	0.21
Despina	53	74	?		0.334655	0.000	0.07
Galatea	62	79	?		0.428745	0.05	?
Larissa	74	96	?		0.554654	0.00139	0.20
Proteus	118	209	?		1.122315	0.0004	0.55
Triton	355	1350	2.14e22	synchronous	5.8768541-R	0.000016	157.345
Nereid	5509	170	?		360.13619	0.7512	27.6

Triton is the coldest in the solar system 38°K and was found a month after Neptune was. Triton orbits against the rotation of Neptune. Nereid is wildly eccentric.

Pluto Satellite	Average Distance (000 km)	Radius (km)	Mass (kg)	Rotation	Days Orbital	Eccentric	Incline
Charon	19,600	593	1.62x10t24	Synchronous	6.38725	?	96.16

Charon is named for the mythological figure who ferried the dead across the River Acheron into Hades (the underworld). (Though officially named for the mythological figure, Charon's discoverer was also naming it in honor of his wife, Charlene. Thus, those in the know pronounce it with the first syllable sounding like 'shard' ("SHAHR en"). Charon was discovered in 1978 by Jim Christy. Prior to that it was thought that Pluto was much larger since the images of Charon and Pluto were blurred together. Charon is unusual in that it is the largest moon with respect to its primary planet in the Solar System (a distinction once held by Earth's Moon). Some prefer to think of Pluto/Charon as a double planet rather than a planet and a moon. Charon's radius is not well known. JPL's value of 586 has an error margin of +/-13, more than two percent. Its mass and density are also poorly known. Pluto and Charon are also unique in that not only does Charon rotate synchronously but Pluto does, too: they both keep the same face toward one another. (This makes the phases of Charon as seen from Pluto very interesting.)

Jupiter's Rings

Ring	Distance (km)	Width (km)	Mass (kg)
Halo	100000	22800	?
Main	122800	6400	1e13
Gossamer	129200	214200	?

Saturn's Rings

Name	Radius Inner	Radius Outer		Approx. Width	Approx. Position	Mass (kg)
D-Ring	67,000	74,500		7,500	(ring)	
Guerin Division						
C-Ring	74,500	92,000		17,500	(ring)	1.1e18
Maxwell Division	87,500	88,000			500	(divide)
B-Ring	92,000	117,500		25,500	(ring)	2.8e19
Cassini Division	115,800	120,600		4,800	(divide)	
Huygens Gap	117,680		(n/a)	285-440		
A-Ring	122,200	136,800		14,600	(ring)	6.2e18
Encke Minima		126,430	129,940		3,500	29-53%
Encke Division	133,580					
F-Ring	140,210			0-500	(ring)	
G-Ring	165,800	173,800		8,000	(ring)	1e7?
E-Ring	180,000	480,000	300,000		(ring)	

Notes:
*distance is kilometers from Saturn's center
*the "Encke Minima" is a slang term used by amateur astronomers, not an official IAU designation.

Uranus' Rings

Ring	Distance (km)	Width (km)
1986U2R	38000	2,500
6	41840	1-3
5	42230	2-3
4	42580	2-3
Alpha	44720	7-12
Beta	45670	7-12
Eta	47190	0-2
Gamma	47630	1-4
Delta	48290	3-9
1986U1R	50020	1-2
Epsilon	51140	20-100

(distance is from Uranus' center to the ring's inner edge)

Neptune's Rings

Ring	Distance (km)	Width (km)	aka
Diffuse	41900	15	1989N3R, Galle
Inner	53200	15	1989N2R, LeVerrier
Plateau	53200	5800	1989N4R, Lassell, Arago
Main	62930	< 50	1989N1R, Adams

Illustrations

Introduction

Solar System Patterns : Image Club Graphics, Circa Art Collection ,CELESTIAL

2-3-4 Patterns: Author

Chapter One

Sun w/Flare: Digital Vision's "Astronomy and Space" collection

Symbol-Sun: Author

Moon, Outer Limits: Digital Vision's "Astronomy and Space" collection

Symbol, Moon: Author

George Bush Chart: AstroDatabank, birth certificate

John Lennon Chart: AstroDatabank, reliable from memory

Asteroid_belt: NASA/JPL Web site

Planetary diameters: Author

SS_Moons: NASA/JPL Web Site

Sun_to_Kuiper: NASA/JPL Web site

Axial Tilt: Wikipedia "common"

Pluto Elliptical: Author

Mercury Elliptical: Author

George Bush Chart: AstroDataBank, birth certificate

John Lennon Chart: AstroDataBank, reliable from memory

Chapter Two

All images by the author, with embedded images of people from Hemera Photo Objects, Vols. I and II

Chapter Three

Moons by Planet-2: Wikipedia "common"
Phobos_sm: NASA/JPL Web site
Deimos_sm: NASA/JPL Web site
Symbol_Deimos: Author
Jupiter-Moons++: NASA/JPL Web site
Symbol-Ganymede: Author
Moons-by-Saturn-2: Wikipedia "common"
Symbol-Titan: Author
UR_moons: NASA/JPL Web site
Symbol_Miranda: Author
Moons by Neptune-2: Wikipedia "common"
Pluto_Charon: NASA/JPL Web site
Symbol_Trion: Author
Symbol_Charon: Author
Symbol_Juno: Author
Symbol_Flores: Author
Ralph Nader Chart: AstrodataBank, considered reliable from memory
Paul Newman Chart: AstrodataBank, birth certificate

Chapter Four
All images by author except:
Planet Matrix: Wikipedia "common"
Planet Pairs: Wikipedia "common"

Chapter Five
Blackboard with Numbers: Hemera Photo Objects, Vols. I and II
Sine Wave: Wikipedia "common"
All other images, base image from NASA/JPL Web Site , added arrows and text by author

Notes
The following collections were purchased by the author and are royalty free:
Digital Vision's "Astronomy & Space" collection – royalty free images
Hemera Photo Objects, Vol I & Vol II -Royalty Free images
Image Club Graphics, Circa Art Collection, CELESTIAL – royalty free images

All charts calculated by the author using author's Intrepid 2.2 software.